W0234701

Reminiscences on Surgery, History and Humanities

Luis H. Toledo-Pereyra
Michigan State University
Kalamazoo Center for Medical Studies
Borgess Research Institute
Western Michigan University
Kalamazoo, Michigan, U.S.A.

CRC Press
Taylor & Francis Group
Boca Raton London New York

CRC Press is an imprint of the
Taylor & Francis Group, an **informa** business

VADEMECUM
Reminiscences on Surgery, History and Humanities

CRC Press
Taylor & Francis Group
6000 Broken Sound Parkway NW, Suite 300
Boca Raton, FL 33487-2742

First issued in paperback 2019

© 2007 by Taylor & Francis Group, LLC
CRC Press is an imprint of Taylor & Francis Group, an Informa business

No claim to original U.S. Government works

ISBN-13: 978-1-57059-697-1 (pbk)

Visit the Taylor & Francis Web site at
http://www.taylorandfrancis.com

and the CRC Press Web site at
http://www.crcpress.com

Cover montage by Kristen Shumaker.

Library of Congress Cataloging-in-Publication Data

Toledo-Pereyra, Luis H.
  Reminiscences on surgery, history, and humanities / Luis H. Toledo-Pereyra.
     p. ; cm. -- (Vademecum)
  Includes bibliographical references.
  ISBN 978-1-57059-697-1
  1. Surgery--History. 2. Medical writing. 3. Surgeons--Biography. I. Title. II. Series.
  [DNLM: 1. Surgery--history--United States. 2. Philosophy, Medical--United States. 3. Surgery--United States--Biography. WO 11 AA1 T649r 2007]
  RD19.T653 2007
  617.09--dc22   2007035023

Dedication

To Marjean, Alex, Courtney, Suzanne and Zach.
For their presence, love and support.

Contents

Section IV. The Date, the Winners

About the Editor

Luis H. Toledo-Pereyra

Surgeon, Researcher and Educator.
He is the author and editor of 20 books.
His books, *Vignettes on Surgery, History and Humanities,*
Origins of the Knife and *A History of American Medicine,*
have been introduced as the standard textbooks for the course
of the History of American Medicine that he has been
teaching for 17 years at Western Michigan University.
He lives in Portage, Michigan,
with his wife Marjean and dog Zorro.

Contributors

Roberto Anaya-Prado
Health Research Division
Hospital of Obstetrics and Gynecology
Mexican Institute of Social Security
Guadalajara, Jalisco, Mexico

Ralph C. Gordon
Michigan State University
Kalamazoo Center for Medical Studies
Western Michigan University
Kalamazoo, Michigan, U.S.A.

Alexander H. Toledo
Department of Surgery, Transplant Division
Northwestern University Medical Center
Chicago, Illinois, USA

Luis H. Toledo-Pereyra
Michigan State University
Kalamazoo Center for Medical Studies
Borgess Research Institute
Western Michigan University
Kalamazoo, Michigan, U.S.A.

Preface

This book follows the same guidelines utilized for *Vignettes on Surgery, History and Humanities* that was recently published. Since the purpose of this work is the same as the *Vignettes* book, the previously published introduction will be used for this preface.

"The importance of medical history in the annals of surgery has been frequently underemphasized. There is so much we can learn from the deeds and examples of our predecessors. There is so much we can admire in their lives and contributions. There is so much we can use to guide our professional development.

This book introduces writings on the history and philosophy of surgery that previously appeared in the *Journal of Investigative Surgery*. These writings were selected and organized after careful analysis to include those works that demonstrated the best cohesive unit in telling about the evolution of surgery and its masters. When necessary we made corrections and added references as appeared to be required.

Our principal objective is to reach the minds and hearts of all students of surgery. This encompasses medical students interested in surgery, surgical residents learning the discipline, faculty surgeons teaching young generations of future specialists, and the practicing surgeons who are making a difference in the community. Additionally, and as importantly, this book attempts to reach students of history in general and those interested in the history and philosophy of surgery in particular.

Our lofty wish is for this book to reach the halls of academia as well as the surgical floors of general hospitals where students, residents, and staff surgeons attend their pre and postoperative patients. Our ultimate desire is that this work will appear during grand rounds and will be a constant companion in the pockets of surgical residents. We further hope that faculty members will frequently test those residents on the extraordinary value of the history of surgery and the wonderful ascent of this prestigious field of medicine.

We hope these writings will present those who read them with encouraging and realistic views of the incredible feats realized by the pioneers of surgery. We welcome new ideas and suggestions the reader might have in improving future editions of this study on the history of surgery."

Luis H. Toledo-Pereyra
Editor

Acknowledgments

The material for this book was previously published in the *Journal of Investigative Surgery*. We appreciate their generous permission to use the *Journal's* content for this publication.

Section I.
Short Notes on Philosophy, Humanities and Writing

CHAPTER 1

Creating the Uncreatable

Luis H. Toledo-Pereyra

A male surgical resident, in his late twenties, could not conceive of making a long-lasting impression in surgery, given its distinguished heritage. He was pensive, lost in complete abstraction, sitting in the surgical intensive care unit where his patient, a 22-year-old female, was suffering from some bleeding after a long pancreaticoduodenal resection.

As his dreams accumulated between thoughts, the resident saw the senior operating surgeon appear at the main entrance of the busy unit. The wondrous specialist, the respected professional, in his late fifties was ready to help. "How can this be?" murmured the resident; most of the senior surgeons do not evaluate surgical care at regular times during the day. Dr. Joe, as he was called fondly by nurses and friends, abruptly interrupted the resident's musings.

"What is up, Axel?" Dr. Joe said.

"We have a case of bleeding in Ms. Ruanto," answered the worried resident.

The savvy surgeon, who had already assessed the case, did not seem impressed by the turn of events. "Sit down, please," said the mature man to his colleague. Immediately thereafter and without pausing, he asked, "What do you know about Tolstoy?"

"Not too much, sir, except that I believe he was a revolutionary who fought in the American Civil War."

"No, Axel, he was a nineteenth century Russian literary giant. Leo Tolstoy (1828-1910) was a celebrated writer, moral thinker and social reformer.[1] Two of his monumental works were *War and Peace* (1869) and *Anna Karenina* (1877), both considered literary classics. Tolstoy believed in the goodness of man. He gave up all his possessions to live with peasants. He insisted on improving the world we live in by loving each other."

"With all due respect, sir, how is this going to help me take care of this patient? Why should I even know about Tolstoy?"

Reminiscences on Surgery, History and Humanities,
edited by Luis H. Toledo-Pereyra. ©2007 Landes Bioscience.

This was an important moment for the sophisticated Dr. Joe, one that he would relish in his interaction with the apt resident. Without appearing condescending, he began, "Dear Axel, you should not look at this event with jaundiced eyes; you should study all great men who went before us and helped build the world in which we live. We owe them respect, and we can learn a great deal from their lives and contributions. So, Axel, as you can surmise, we have great knowledge to acquire before being considered citizens of this world. Before we are given the responsibility of taking care of patients, I think we must possess the necessary knowledge of the human race we are entrusted to care for. I asked myself, how can we care for other fellow humans when we have not learned how others have lived and suffered in the past? Tolstoy can show us a portion of the answer. This is why, I think, we need to engage ourselves in advancing our dedicated knowledge about others. *Anna Karenina*, for example, is a novel of love, betrayal, social conflict, and personal moral principles. All these characteristics make the world become more human, in a way that medicine becomes more human for the practicing physician. A. N. Wilson, a distinguished Tolstoy scholar, summarized the master's presence in the following manner:[2]

> *He stood for something much bigger and more important than just himself or his ideas. So long as he was there, huge numbers of Russians felt that it was not quite impossible to believe in the prospect of individual liberty, the survival of individual dignity in the face of cruel, faceless, bureaucratic tyranny.*

"You see, Axel, you can be a good doctor or surgeon without world knowledge, but to be great and aspire to greatness, to be able to create the uncreatable, to advance world experience, we need to study deeply the contributions and behaviors of others, particularly those who paved the way for incredible developments in the past. Creating the uncreatable is reaffirming the generations of individuals who changed the world, is conceiving life as a temporary passage that offers the possibility of change, is to challenge the problem and encounter the ideal solution, is to live fully aware of our forefathers and their significant contributions, is to appreciate life and all its complexities."

Dr. Axel, who had attended one of the best colleges in the nation and had received an excellent scientific and technical education with minimal emphasis on humanities, had considered himself well prepared prior to Dr. Joe's argument. It was evident now that there was a great deal yet to learn.

Axel calmly asked his surgical mentor, "What about the care of this sick woman?"

Dr. Joe, without great misgivings, quickly responded. "I think this lady does not have too much of a problem since the rate of bleeding is slow and quite usual after this long and extended surgery. I would continue her observation, obtain periodical hemoglobins, and check her coagulation parameters."

At this stage, the young surgeon understood the value of the lesson. First, he needed to review the pathophysiology of care for each patient. Second, and just as importantly, he needed to grow in human knowledge to be worthy of caring for his patients. Dr. Joe had certainly impressed this rising surgeon and left an indelible message: to create the uncreatable one needed to learn from the past, the common people, and the great contributors of our time; to be loyal to the annals of the history of the world; to learn and assimilate what others tell us with their voice, their writing, and their principles; and to be ready to advance the cause of many suffering from disease, family difficulties, and economic despair. "Help others at all times," continued the older surgeon, "since creating the uncreatable is not only desirable in science and medical advances, but in respect, love, and concern for other human beings." It was clear that Dr. Joe had made his case. At the end of the day, both aspiring surgeon and patient had improved and were better for it.

References

1. Tolstoy, Leo. World Encyclopedia. New York: McGraw Hill, 1969.
2. Wilson AN. Tolstoy. New York: Ballantine Books, 1988:6.

CHAPTER 2

Welcoming a New Hip

Luis H. Toledo-Pereyra

Facing his wife's major operation, the senior surgeon found a small amount of comfort as he paced the surgical corridors near midnight. He paused for a brief moment to reminisce. He clearly pictured the good days of college, he remembered his early years in medical school, he recalled the arduous cases he encountered during his surgical residency, and most importantly, he remembered his family and especially his wife, whose condition worried him continually. She had been given a direct verdict—You need a new hip! Overwhelmed, he thought about the complications of the procedure and the perils of life with an artificial hip, particularly for someone only 56 years old. What could they expect? Certainly, a new life. A life with a new hip. What a change. One that, in spite of his medical knowledge, he did not know how to deal with. It was like patients asking him, "*What would you do if it were your wife?*" "*I would welcome the new hip into the family,*" abruptly said the pensive surgeon.

As judgment day neared, the tension in their lives was evident. There was no place for a surgeon. Instead, his wife needed a loving— though preoccupied—husband. The senior surgeon had to set aside his usual role, to view events from the perspective of his wife, a suffering hip patient. What was she going through and how was she coping? Extremely well, he thought. As the preoperative visit ended, the orthopaedic specialist noticed certain nervousness in the couple. A mild antianxiety drug, Xanax™, was recommended, and the day for surgery was set. A day that was not going to be soon forgotten!

The husband surgeon recorded some of his personal deliberations.

The Suffering Hip: A Long Road

How hard would it be to have a surgical hip?
How easy would it be to tolerate the increasing pain?
How hard would it be to become a monk?
How easy would it be to be agnostic?

Reminiscences on Surgery, History and Humanities,
edited by Luis H. Toledo-Pereyra. ©2007 Landes Bioscience.

A hip has its own life
Works all the time until midnight
Walking or running do not faze her
Until something intrudes-the arthritic element

Living with pain is not fun
We talk about it without remedy on hand
Do elephants have painful hips?
How do they solve their immense problems?
Do we give them a new hip also?

Pain comes from osteoarthritis
Pain pills do not always work
Exercise frequently does not help
And the only friend is total hip replacement

Hips are not of bone only.
Hips can cry too.
Hips have feelings
Could you walk only on vitamins?

Technology has made wonders
Look at Charnley and the steel and plastic hip
Look at Thomas Gluck in Germany in 1890
Look at all great hip innovators
But, can I sleep on my side without pain?

The surgery date—August 16—came quickly, sooner than anticipated. Never mind about not being able to sleep. Five in the morning as though on a normal day, driving to the hospital, being admitted and taken to surgery, all was so real that it was surreal. The husband and his daughter said good-bye to their loving wife and caring mother. At 8:00 a.m., the waiting room clerk made it official, the orthopaedic surgeon had opened, exposing the patient's hip and encountering her soul, a time in which both patient and family exist outside time with no control and no immediate way of return. Two hours later, husband and daughter were informed that the operation was advancing as planned. At three hours, they began to inquire and the response was that unexpected findings required further revision and reconstruction. This was not an anticipated response and was definitely unsettling. Having been on the operating side, the senior surgeon knew what those remarks might mean. He too had used them to explain delays caused by critical and difficult operative problems. There was nothing to do but wait with hope and optimism.

After three and a half hours, the orthopaedic surgeon appeared, he looked worn out by the immensity of the case and said, "*She is doing fine even though some unexpected findings were encountered on the socket of the hip. It was worn out further than could be recognized preoperatively. It required more surgery, reconstruction, and a bone graft in addition to the replaced hip. She will not be able to bear any weight for the next two months.*" This was the final verdict. She was out of the operating room and beginning her immediate recovery. Only later did the husband discover that she had multiple bouts of cardiac rhythm abnormalities (with increased risk of further problems) during surgery and in the recovery room.

Four hours later, husband and daughter saw her in her room for the first time. She looked pale, withdrawn, but with the usual sparkling, high intellect, asked, "*Is my leg still here?*" "*Yes, of course,*" they said. There was no question that recovery was going to be long, tedious, committed, and filled with surprises.

Without delay, on the first day after surgery, hip camp began. Strenuous exercises were oriented at reactivating dormant muscles of the thigh. A painful process since the traumatic effects of surgery were still evident and would linger for several weeks or months. Narcotics flowed steadily since pain was the main oppressor. As the days passed, it was clear that recovery was an arduous and complicated enterprise. Learning how to sit, how to attend to her most personal necessities, how to turn and prevent flexing of the hip less than 90 degrees were tasks that preoccupied the husband surgeon. Her care demanded continuous and careful attention to detail.

The extraordinary potential risks and benefits of total hip replacement (THR) sent the husband surgeon in search of historical details. How had surgeons thought about reaching the well hidden hip joint? Whose idea was replacing the hip? Why did it take so long to help so many arthritic patients? No clear answers were forthcoming, and the scientific development of modern hip arthroplasty seemed diverted by detours. Two of many surgeons were credited with major developments: in 1923, Marius Smith-Petersen from the Massachusetts General Hospital in Boston,[1] and in 1958, John Charnley from Wrightington Hospital in Lancashire, England.[2] Smith-Petersen performed the first THR operation on the American continent, but unfortunately failed when using a glass cup to contain the femoral head. Charnley thoroughly succeeded with a steel femoral component and a plastic socket. New advances in biomechanics, surface replacement, and biological response are being introduced now with positive outcomes. Still many questions remain unanswered. What is the best hip replacement for the young?

What are the toxic effects of metallic ions? Can ceramics increase prosthetic endurance? What effect does wear debris have on the body? In spite of these pressing but unresolved issues, THR helps thousands upon thousands of patients every year.

At home four days after surgery, his wife was far from resuming normal life. The husband surgeon was confused and overwhelmed by the difficult situation. The pain was severe, persistent, and only the powerful Vicodin™ could control it periodically. Getting around was an ordeal and using the bathroom as well. The limitation of hip movements and not knowing exactly what was good for her were at the center of the discomfort equation. Now, sleeping was not a simple issue since the position of a recently operated hip (with a fresh 30-cm incision) did not constitute a Sunday afternoon picnic. For this patient and her care giver discomfort remained eminently high. Controlled pain and movement would require more time, more healing. Hadn't this operation been performed to alleviate pain?

Two weeks later, at the first postoperative visit, the initial concerns about early bleeding, respiratory ailments, wound infection, and nerve damage from surgery were dissipating, and a successful early outcome seemed secure. When his wife made an appointment to have her nails done on September 2 (three weeks after surgery), the husband surgeon knew the initial recovery was progressing at a good pace, even though pain still clouded many activities. One month after surgery, things appeared calm, but taking full steps on the operated hip was not routine yet. At least pain was less prominent and graduation from Vicodin™ appeared reachable. Husband and wife were confident that this unique and memorable surgery would bring better walking, considerably less pain, and opportunities for a full and rejuvenated life.

What did this husband and surgeon learn from this important and significant experience? I learned something that perhaps I already knew but have frequently forgotten—that life is unpredictable and many times uncontrollable, that life can bring circumstances not readily explainable, and that God—although others call it luck—can take an important part in the main event as well as in the recovery process. I also learned that it is difficult to care for your relatives, particularly your wife, and that close and continuous acute care could become very stressful. Myriad feelings crowd my recollections and accompany each day's victories and challenges. I was exhausted but profoundly understanding of the courage and determination of my wife in this most overwhelming and extensive operation. As we start another month of recuperation, it is clear that the future is still not completely defined. More months of exercise remain ahead. More experiences are to be encountered, and more life is ready for us to live!

References

1. Smith-Peterson MN. Evolution of mold arthroplasty. J Bone Joint Surg 1949; 30B:59-75.
2. Charnley J. Evolution of total hip replacement. Ann Chir Gynecol 1982; 71:103.

CHAPTER 3

Working Less and Being Happy

Luis H. Toledo-Pereyra

It is a frequent and almost constant desire of our society to work relentlessly for as long as possible so we can accumulate good deeds of continuity and perseverance. What is it with all of us that we want to endure a difficult and over-taxed existence in this world? When should we pursue different and more humane goals? I would like to advance a new perspective and priority to my surgical colleagues. What if we could work less and be proud of it? What if we could reach happiness with a tamed and moderate life? What if we could maximize our goals with a simple existence? I do not think I would see myself uttering these challenges a few years back, even though today, I can judge them to be completely appropriate and worthy of special consideration.

Where do we begin the awesome task of challenging and analyzing our current societal aspirations? Where did the notion of working hard come from? Where did the idea of arduous and persistent labor originate, and why is it so frequently associated with success and job advancement? Is it because the more we work, the more we are seen as dedicated and accomplishing a great deal? I do not have the right answers for all these important questions but allow me to start with a few considerations. It is not difficult to figure that since very early times the intellect of our human ancestors began to deal with ideas of work, laziness, advancement and good life. It is obvious as well that dedication was seen as synonymous with accomplishment and success. Professionals that worked hard were seen as reaching higher levels of stardom that those who took life in a calmer and slower fashion. These ideas were perpetuated in the evolution of humanity and clearly conveyed that hard work was the only means towards pursuing a valuable existence.

Is it possible to conceive that we can work less, accomplish more and therefore be happy? I believe we can but we need to work smart. Working hard is not enough, working smart is. What I mean is the

---

Reminiscences on Surgery, History and Humanities, edited by Luis H. Toledo-Pereyra. ©2007 Landes Bioscience.

idea that I recently heard on Neal Bortz's radio show that working smart is the ability of each individual to use his/her intellect, knowledge and capacity to structure his/her own case. The great Renaissance man Leonardo da Vinci attempted to persuade others that "the greatest geniuses sometimes accomplish more then they work hard".[1] One could paraphrase with others on this statement by indicating "the greatest geniuses sometimes accomplish more when they work less." The greatest challenge would certainly be to convince higher authorities that time spent in reflection, contemplation or meditation, but not necessarily doing the hourly labor, would translate into a more efficient and accomplished job. "Great musicians claim that their music comes to life because of the spaces between the notes. Similarly, the spaces between your hardest working efforts allow ideas and solutions to incubate and grow".[1]

It appears then that three important elements are at play in this writing, hard work, success and happiness. Hard work does not need definition, success and happiness do. Society has not infrequently taught us that success is dependent on financial remuneration and personal recognition. These are two characteristics fully inclusive of a successful life but not necessary to reach this coveted prize of society. Success should be seen in relative terms since it is a very personal accomplishment. Success is intimately associated with happiness. Success without happiness is not meaningful. Happiness without success can be realized but requires a great deal of personal belief. At the same time, happiness is a sense of satisfaction and pleasure and since success can give you satisfaction and pleasure also, both are closely associated and are an intricate part of each other. Robert Updegraft said that "happiness is to be found along the way, not at the end of the road, for then the journey is over and it is too late".[2] These words clearly reflect the significance of enjoying the moment of success, the importance of living with happiness along the way and the belief of searching for better means of finding the fountain of success and happiness. Kate Wolf would summarize her characterization of success with the following words, "find what you really care about and live a life that shows it".[3]

How would all of this apply to the career and life of a surgeon? Indeed, it applies as much or more than it does to anyone else. Let me introduce to you what you already know. As two surgical residents were meeting in the surgeon's lounge, it was evident that hard work was on their minds. They had been on call and working for 30 hours without stop. They did not know how to think straight at this point.

However, they began their conversation in the following way, "how can we do the same things without having to work so excessively and being at the hospital all the time?" one of them asked. The other one, who was a senior resident, referred to the tradition built by the surgical masters of the past. "How can we change this tradition and introduce principles of moderation in working hours and surgical practice?" the junior resident questioned. They did not know how to respond to the puzzling situation that was before them. A real quagmire created since modern times of organized surgical residencies in the twentieth century. The motto being, "the more you work, the more endurance is built and the more suited you become for the surgeon's life." This is a principle that has been adhered to by surgical residents of past and present generations, including mine; a principle engrained in the minds and bodies of all aspiring surgeons, a principle that did not require explanation and a heritage all surgeons were proud of. To change the minds and beliefs of the surgical world would be an insurmountable task.

Everything began with William Halsted (1852-1922) who created the first surgical residency in Baltimore. He established a system that required maximal dedication, attention to detail, complete endurance and an open-ended time of residency completion. This system started to produce well-trained surgeons who had spent uncountable hours of committed work in the surgical suites and operating room. Other surgical teachers in different parts of the United States and the world implemented a similar successful system of hard work and commitment. The modern era of surgical residency started with principles intimately adhered to Halsted's ideas and teaching concepts of hard work and dedication. How can anyone reorganize this system by claiming that less work would create a successful surgeon with a happier life? It had not been tried for 100 years until very recently, in 2003, when the Accreditation Council for Graduate Medical Education (ACGME) implemented a challenge to surgeons and the rest of the medical profession. It required less hours of work for residents (80 hours/week maximum) while maintaining them productive and well educated in their careers. A new and different scene was presented to surgeons with no opportunity to regress.[4-7]

Years before, in 1984, as a consequence of the death of Libby Zion at New York Hospital, attributed by the prosecution in great part due to resident fatigue, the Department of Health of New York State began to face the introduction of resident's work hour restrictions.[4]

Other programs were to follow this example in California and other locations.[5]

A great number of papers have recently appeared in the surgical and medical journals evaluating the new system of less working hours in surgical residencies.[5-7] The preliminary results appear to indicate that there is no harm to patients or to the education of residents by restricting the time of work. Now, the second part of the equation has not been completely answered: are surgical residents happier than before? The initial response is positive but a valid answer will require more years of careful study of the new system. The majority of the surgical residents appears to be satisfied with less hours of work,[5-7] and continues to be productive in their tasks.

After finishing the surgical residency, let's examine the life of the practicing surgeon. Here, our contention that you do not have to work that hard to be successful and entirely happy might be more real. It is clear that the surgeon in practice could limit in many ways his time dedicated to patient care in urban areas, outside of academia and in small practice groups. The time could be eventually managed individually to satisfy personal and family goals. Success and happiness then, would be a more personal endeavor.

Another important question remains, what about the surgeons working in academia and/or large surgical groups? This is an entirely different scene since other factors will regulate the surgeon's practice. In this case, time of coverage might be better accomplished with less working hours but increased commitment and less freedom at the time of regular shifts will be evident. Here, again we are dependent on the style and aspirations of the particular surgeon's life. For some, working less would not be an option since this concept was not in their plans. For others, this idea would yield positive emotions, and they would welcome the possibility of change.

The initial proposition of less work, success and happiness is still a good possibility in progress. In this regard, I believe, that the time allowed for critical thinking and/or relaxation instead of continuous work would be helpful in increasing the overall accomplishments of that particular individual. It would be fair to say that the effects of less work, success and happiness will be dependent on the individual person pursuing his/her own special aspirations. From my own perspective, I would support the notion that it is possible to reach success, and be happy without working excessively hard. It is evident perhaps, that the quality of well-structure work, even limited in time, would offer a better alternative than the quantity of work performed for quantity sake. I am optimistic that the concept of less work, more accomplishments and happiness, will stimulate others to consider that it is feasible to reach this status in life without having to apologize for reduced amount of work. I would leave the rest of the conclusions to your own judgment, hoping that you will review the concepts presented herein within the whole perspective of your unique life and future aspirations. Just remember, that hard work does not warrant success, and that you could obtain as much happiness and reach your expected accomplishments with less but efficient work.

References
1. http://www.48days.ibelieve.com.
2. http://www.innerself.com.
3. http://www.bruceelkin.com.
4. Asch DA, Parker RM. The Libby Zion case. N Engl J Med 1988; 312:771-775.
5. Schwartz RJ, Dubrow TJ, Rono RA et al. Guidelines for surgical residency working hours. Intent vs reality. Arch Surg 1992; 127.
6. http://www.entlink.net.
7. Brennan TA, Zinner MJ. Residents' work hours: A wake up call? Int J Qual Health Care 2003; 15:107-108.

CHAPTER 4

Mourning the Prostate

Luis H. Toledo-Pereyra

The message was clear and unambiguous, the urologist had given his unmistakable dictum: "You have prostate cancer." "How is that possible," replied the immediately concerned 61-year-old senior transplant surgeon, "especially when there were no signs of a tumor with digital rectal examination and the ultrasound was completely negative?" The prostate specialist reorganized his thoughts before answering and said, "I never thought you had a good chance of having a positive biopsy either. The odds were in your favor." Without pausing, the urological surgeon added, "It is hard to understand these results, but now you need to remember these figures. Write them in some secure place. You are *T1c, Gleason 6, PSA 4.1*. Never forget these figures, they will give you a prognostic indicator for the future. You need to become fully aware of their meaning." The patient did not know what the urologist meant. Even though he was a surgeon, a transplant surgeon, his knowledge of the prostate gland was primitive at best. He needed to go back to the library to learn the fundamentals of this disease.[1-7] It was clear that the burden of inquiry was on his side. A decision was required as to the optimal individual mode of treatment, the best that could fit his way of life and thinking, not an easy task at this early stage of diagnosis.

After uncountable visits to the library and discussions with judicious medical colleagues, it did not appear that radioactive seeds, Mayor Giuliani's choice, were appropriate for an enlarged prostate of 120 grams plus. The possibility of brachytherapy was not prudent in this case either. Three options were left: surgery, regular radiation therapy, or watchful waiting. Since the last one did not represent a truly sound possibility in this case and because radiation therapy represented a long-term approach of six weeks of daily treatments, the surgeon-patient decided to proceed with surgery. Now, what would be the best surgical procedure? This was a critical and defining ques-

Reminiscences on Surgery, History and Humanities,
edited by Luis H. Toledo-Pereyra. ©2007 Landes Bioscience.

tion that needed ample and detailed analysis.[1-7] No rushing allowed under any circumstances.

After more weeks of literature review and internal searching, he favored the removal of the prostate by laparoscopic techniques,[6,7] a novel approach he considered to be less invasive than the regular approach, either retropubic or perineal. He pondered the potential advantages of less pain, smaller incisions, less blood loss, and earlier hospital discharge with the laparoscopic procedure. He embraced history as a safe and protective haven for understanding the evolution of the various procedures and important tests in the management of patients whose best option was surgery. He could not believe that even though urological procedures were old, techniques to operate on the prostate were basically in their infancy, and most of them were essentially from the twentieth century.[3-7] Goodfellow (1891), Young (1904), and Millin (1947) were the early pioneers in understanding this hidden gland.[3] As incredible as it might be, testing for the prostatic surface antigen (PSA) was just introduced in 1986, and the ultrasound-guided biopsy two years later.[3] Not until 10 years thereafter, in 1997 and 1998, did the laparoscopic prostatectomy arrive.[6,7] As a surgeon, he realized he was dealing with the early stages of a new, very new technique. Consequently, this treatment option required the ideal place and best surgeon, as only a few surgeons were performing laparoscopic prostatectomy on a routine basis.[1-3,6-7]

Further investigation led to establishing a good relationship with a noted prostate cancer surgeon in Florida who had brought the procedure to the region several years back. After speaking with former patients and clearing medical insurance issues, the operation was scheduled for January 5, 2005. There was no going back. Instructions had been given, approvals were circulated, and permissions had been reached with employers. Now the date was real and could not readily be erased. The operation was set!

The surgeon-patient needed to put his mind and emotions in order. "What would this procedure do to his intimate life?" he wondered. "What would be its effect on sphincter control and other side symptoms?" "Would he be able to tolerate anesthesia since this was the first time, at 61, that he was undergoing general anesthesia?" A great number of questions arose from his inquisitive spirit. The answer was unmistakable: "The tumor had to come out! There was no other way out," he figured.

After a long series of presurgical tests, including an unending and primitive stress test, the word was out. "You can tolerate surgery!"

said the internist. Wife, suitcases, and aspiring thoughts were packed together to travel towards recovery with serious but high hopes. The couple considered general anesthesia and its potential effects, the possible and perhaps unexpected findings of surgery, and hoped fervently that the tumor would be confined to the gland.

Promptly upon their arrival, they met surgeon, anesthesiologist, and surgical staff. The day of surgery arrived, and all doubts were erased with surgery imminent. As Versed™, a surgical anesthetic agent, was administered, the many questions seemed unimportant and life was suddenly simpler. Exactly two hours and fifty minutes later, the prostate specialist appeared in the waiting lounge. "The prostate was monstrous in size, and all was removed. Now we need to wait for the pathology report," he confidently reported to the worried wife.

Meanwhile, the surgeon-patient regained consciousness in the recovery room and began dealing with new questions. "Is everything done? Is the surgery finished?" he asked. Even though both were true, he felt that it could not be done that promptly and efficiently. Some abdominal pain, several surgical scars, and a well-secured Foley catheter were the main vestiges of the surgery. The well-proclaimed foe, an enlarged and diseased prostate, was by now in the hands of the knowledgeable and able pathologist.

An uneventful night passed, surgeon-patient and wife had no complaints, and at midday the distinguished urological surgeon appeared at the entrance of the room, moved toward the patient, and swiftly removed the pelvic drainage, felt the wounds, and declared the initial battle over. The patient could leave. "See me tomorrow at 10 a.m.," he said as he stood in the doorway. There must be other patients requiring his attention, we thought.

The next encounter with the reputed urologist was crucial. The pathology report was at the forefront. Nervousness permeated the room, as thick as a heavy fog. The specialist suddenly appeared in regular street clothes, no starched shirts or accompanying ties, and said, "Didn't I give you the pathology report?" "Not really," said the surprised and extremely anxious surgeon-patient. "Very well. You did good. The surgical margins were negative. You are now *T2c, Gleason 7, negative margins, prostate of 135 grams*. These are your new statistics," exclaimed the pleased prostate surgeon. And then he added, "You have a good chance of living another thirty years." No good photographer or gifted artist could have depicted the aura of happiness on the faces of the surgeon-patient and wife. Even though he did not know exactly what T2c meant, he figured that Gleason 7 went up a notch from the initial Gleason 6, and T2c, he deduced after reading the literature,[1,2] meant that the tumor was present in two lobes of the prostate without extracapsular involvement. "The surgical margins were free and that was what mattered most," he mused to himself, almost incredulous.

Being a religiously committed individual, the surgeon patient acknowledged the helpful hand of God in these events. He personally thanked God for enough strength to face this opportunity and for the initial positive result. The rest, Foley catheter still in place for ten days, unknown length of urinary incontinence, and important matters of intimate discourse were secondary after hearing the pathological report. Everything else seemed so miniscule at this point!

Exactly three days after surgery, when the surgeonpatient and wife were flying back from Florida, hundreds of thoughts inundated their minds, but the most fundamental of all was, "The tumor is confined to the surgical margins. They are free." This sentence clearly summarized his experience and reverberated throughout his mind. Nothing else needed to be said. As the plane took off, husband and wife said their prayers and thanked God for this incredibly positive experience.

Two important questions remain: Why is this writing so special for the surgeon-scientist? How can this experience have an effect in the professional life or human endeavor of the practicing surgeon? Undoubtedly, the personal experience of a fellow surgeon enhances our perspective of the profession and, equally important, increases our experience as human beings. Recognizing how others handled a significant personal event, such as the one described here, can give us a better understanding of the important human side inherent in the surgeon's private life. It also allows us to contemplate how unpredictable events might improve our lives within the context of this experience. And, finally, the positive assimilation of experiences-our own and others'-permits us to be better human beings and, in the end, better professionals.

References

1. Marks S. Prostate and Cancer. 3rd ed. Cambridge: De Capo Press, 2003.
2. Walsh PC, Worthington JF. Guide for Surviving Prostate Cancer. New York: Warner Books, 2001.
3. Brosman SA. Radical Prostatectomy. Prostate Cancer Research Institute, 2003, (Available at: http://www.prostatecancer.org).

4. Young HH. Early diagnosis and radical cure of carcinoma of the prostate. Bull Johns Hopkins Hospital, 1905:175:315-321.
5. Millin T, Dubl MC. Retropubic prostatectomy: A new extravesical technique. Lancet 1945; i:693-696.
6. Schuessler WW, Shulam PG, Clayman RV et al. Laparoscopic radical prostatectomy: Initial short term experience. Urology 1997; 50:854-857.
7. Guillonneau B, Cathelineau X, Barret E et al. Prostatectomy radicale coelioscopique: Premiere evaluation apres 28 interventions. Presse Med 1998; 27:1570-1574.

CHAPTER 5

Slow Down for Heaven's Sake

Luis H. Toledo-Pereyra

Since time immemorial, surgeons' reputations have been built on the presumed necessity of working quickly. This is complete nonsense, and let me tell you why. The mandate for speed originated from the times of no general anesthesia and poor analgesia. Speed was considered essential in the times of continuous suffering and disrespect for tissue and organ function. These were the times of scientific disregard, when practicality ruled in the absence of a firm theoretical foundation. Under these precarious conditions, the dictum was: "*Operate fast regardless of results.*"

With more than one hundred years since the discovery of anesthesia (1846), and with many incredible surgical developments at hand, the speed of surgery should no longer be an issue. In practice, speed is so ingrained in the surgeon's mind that many surgical teachers continue to praise an outdated pace as a surgical virtue. Why? Do patients who undergo a surgical procedure with five hours and twenty minutes of anesthesia fare better than those whose surgery lasts four hours and fifteen minutes? If there are no surgical complications during either case, clearly the answer would be no. Surgeons are aware of this, of course, but many continue to act as if speed still matters. So when will the need for speed disappear? Must we turn to physiologists and anesthesiologists for arguments that will sway today's illustrious cadre of surgeons? Or can we relax and allow our resident colleagues or future surgeons to operate calmly and without the nagging pressure of haste?

Consider a recent hypothetical example. At a respected Midwestern regional university hospital, all operating rooms were empty. Surgeons and assistants had finished their cases by late afternoon and the nurses and technicians were preparing for the next day. The hospital had developed strict guidelines about room turnover, operating times, and complete efficiency. Everyone had to comply by the clock, strictly speaking.

Reminiscences on Surgery, History and Humanities,
edited by Luis H. Toledo-Pereyra. ©2007 Landes Bioscience.

As the head nurse, Elizabeth Svensen, a mature and extremely conscientious lady, was ready to leave the premises, she noted that the lights were still on in Room 14, as if somebody were just beginning the day's work on that cold and misty February evening. She walked to the room, slowly opened the door, and realized that Dr. Joe, as he was affectionately known, and the senior surgical resident, Dr. Ed Macky, were reviewing surgical steps for the next day's case. Dr. Joe was talking through each one of the important hurdles of the procedure, a parotidectomy on a 65-year-old patient with parotid gland cancer. Dr. Joe was showing Dr. Macky the instruments, the various tissue planes, the preservation of the nerves, and the best way to enucleate the gland while leaving plenty of margins.

Ms. Svensen asked, "What are you doing here so late?" Dr. Joe, as the faculty surgeon, said, "We are rehearsing the surgery for tomorrow. It will be a complicated one, and I want Eddie to do well." Ms. Svensen admired the doctors' dedication, but still could not comprehend why they were practicing in the OR. Why not use the classroom?

Dr. Joe recognized the bewildered look and rushed to explain. "Elizabeth," he said, "I hope we are not delaying room preparation. We're here because Dr. Macky has advanced extremely well in his residency, but has been frequently criticized for his deliberate slowness in performing surgeries. In fact, he has been threatened with probation if he does not improve his speed." Gaining momentum, Dr. Joe continued, "We're here to offer Eddie the real arena in which he will be performing before a group of senior surgeons. Dr. Richard Versad—the most senile, dogmatic, and unfair of them, who acts as if he would have been trained by American Civil War surgeons—will be evaluating Eddie's speed in the parotid tumor case tomorrow. I never heard such an irrational approach to the training of surgeons...otherwise ask the master of modern American surgery, William Halsted."

Dr. Joe was himself a firm follower of William Stewart Halsted's (1852-1922) ideas and principles.[1] When Professor Halsted introduced the great benefits of the 'surgery of safety,' he changed modern surgery forever. Halsted, the surgical scholar for excellence, convinced all his disciple surgeons that speed was not the foundation of surgery. Like many other distinguished surgeons who followed his school, Halsted taught the importance of using time wisely, of deliberately focusing on safety, of continuously protecting tissue from the roughness imposed by time constraints.

"I see," said Ms. Svensen. "So I should be more consistent in protecting residents from the tyranny of time. I realize now what you've so frequently commented, Dr. Joe. *You do not have to be that fast in surgery to be good, otherwise ask William Halsted.* Is it not so, Dr. Joe?" "Indeed," replied Dr. Joe, "and thank you, Elizabeth, for your kind understanding."

The famed Tulane surgical professor, Rudolph Matas (1860-1957), appreciated Halsted's contributions to careful surgery in an article written after his death.[2] Matas defined brilliance not based on ill conceived speed, but on the ability of the operator to demonstrate good judgment during surgery. In Matas' own words:

"Allow me to detain you for a brief space with a few reflections, suggested by the commentary, occasionally heard, that Dr. Halsted was not what is popularly described as a 'brilliant operator,' a statement which might be interpreted as depreciatory of his technical abilities by those who are unfamiliar with his aims as a surgeon and the principles that governed his operative acts. If by 'brilliant' we mean the surgeon who utilizes his opportunities to dazzle the public with the prodigies of his skill, who listens for the plaudits of the multitude more intently than he does to the murmured approval of his conscience, and who burns his incense to the gods of the gallery, then, we must agree, Dr. Halsted was not one of that class. But, what do we mean by a brilliant operator? In the sense in which it is most commonly used, brilliancy is a quality whose chief characteristic is speed, the quickness and dexterity with which an operator executes and accomplishes the operative act. This is a quality in which our forefathers excelled, to acquire which they bent all their energies, and in which they vastly surpassed us. In this respect, we, the surgeons of the present generation, can no more compare our performances with theirs than we can make comparison between the speed of a horse car and that of a twentieth century limited railroad express. But when we consider the effects of a collision between horse cars, on the one hand, and railroad trains, on the other, including the wreckage that follows in each case, we may form some idea of the relative effects of speed as applied in the cyclonic operations of the older surgery and the calm but sure and safer motions of the surgery of the present. Happily for us and for humanity, the time has long passed when surgical brilliancy and ability

could be gauged by the clock, or when the relative merits of surgeons could be estimated by the rules of the prize ring or the authority of the Marquess of Queensbury. That was well enough in the dim days of antiquity, in the days of Galen and Celsus, when limbs of conscious men were amputated with an axe or a guillotine; or in much later days, when a Lisfranc, a Dieffenbach, a Lizars, or a Liston, could disarticulate a hip in five minutes or less, provided that in the flourish of blades, one or more of the assistants were not put hors de combat by the lightning maneuvers of the operator; or that one could say of a modern master what was said of Fergusson, who, in lithotomy, proceeded with such lightning speed and skill that someone advised a prospective visitor to his clinic to, 'Look out sharp, for if you only wink you will miss the operation altogether!' "[2]

No doubt Matas exhaustively clarified this issue for us and for the incredulous surgeons who continue to believe that speed is the essence of surgery.[3-5]

The next day Dr. Macky was undergoing his evaluation by the stern, mister-no-personality Professor Richard Versad, and things, for a change, were going well except for the final nerve separation and tissue closure. Dr. Versad uttered a lament, "Residents never learn how to hold the needle holder and how to wipe the blood oozing from injured tissue." No response was expected, and therefore, the comment went unanswered. As Dr. Versad was leaving the room, he said, "Very well, Macky, I feel better about you, but you are still lacking some speed even though your patient did satisfactorily." "Thank you," Dr. Macky responded. He knew, however, that Dr. Versad was not going to understand that one could slow down and still be good.

"Different schools of thought, I guess," said Dr. Joe amicably as he was leaving the surgical dressing room. "A long and fruitful day," said the pleased and perplexed resident. Even after centuries, it would take more time for most surgeons to recognize that it is possible to *slow down and still be good in surgery, otherwise ask William Halsted.*

References

1. Toledo-Pereyra LH, William S. Halsted. Cirujano, Maestro e Innovador in Maestros de la Cirugia Moderna. Ed. Fondo de Cultura Economica, 1996:87-110.
2. Matas R. William Stewart Halsted: An Appreciation. Johns Hopkins Hosp Bull 1925; 36:2, (1852-1922).

3. Burke WC. Surgical Papers of William Stewart Halsted, Vol. I. Baltimore: Johns Hopkins Press, 1924.

4. Cushing H. William Stewart Halsted. 1825-1922. Science 1922; 41:461.

5. Nuland SB. Medical Science Comes to America: William Stewart Halsted of Johns Hopkins in Doctors: The Biography of Medicine. New York: Alfred A Knopf, 1988.

5

CHAPTER 6

The Surgeon as a Scientific Writer

Roberto Anaya-Prado, Alexander H. Toledo and Luis H. Toledo-Pereyra

Introduction

As Booth eloquently stated, "When standing in the reading room of a library, one can see around centuries of research, the work of tens of thousands of researchers who have thought hard about countless questions and problems, gathered information, devised answers and solutions, and then shared them with others".[1] A medical or surgical library offers the same unique environment. Surgeon-teachers dedicate themselves fully to research, to encounter better ways to improve their instructional skills, to hone the new advances of the discipline, and to discover old or new pathways of excellence. Research constitutes the life of the academic surgeon and scientific writing the most important endeavor. Surgeons without the ability to convey their experiences cannot effectively participate in the advancement of their discipline. It is our responsibility to enhance and teach the virtues of scientific writing. Those in the surgical profession who never learn how to write miss the opportunity to explore new scientific frontiers. In this writing, we address some of the most frequently cited remarks, frustrations, and problems regarding surgical and medical scientific writing.

Why Surgical Writing Is So Difficult

Much has been written and spoken about the low level of scientific writing, often with good reason.[1-14] If a piece of surgical writing is hard to understand, usually it is the shortcoming of the author not the reader. A journal, textbook chapter, or abstract that is clear, concise, and well organized is easy to read. The good surgeon-writer uses short sentences, simple and concrete words, transition words, and

Reminiscences on Surgery, History and Humanities,
edited by Luis H. Toledo-Pereyra. ©2007 Landes Bioscience.

has a logical sequence of ideas. Writing that is clear, concise, and well organized is not an accident. It's deliberate and hard work.

How is it that surgeons who are frequently brilliant in one area are so lacking in another? Writing is a deliberate endeavor that requires experience and attention to technique. The best writers spent time honing their skill. This skill is still a prerequisite for academic surgeons. In fact, many times the only thing a person knows about surgeons or physicians is what they put on paper. For that reason, writing should be a priority. Writing is more skill than talent. Of course, there are those who like and those who do not like to write. Learning to write in the second case becomes a problem. Nonetheless, any surgeon or physician can and should learn to write effectively.

Effective surgical or medical writing requires the same qualities of thought that are needed for the rest of science: logic, clarity, organization, and precision. Passion for one's niche is also helpful as surgical writing is difficult when dedication and interest are missing. Surgical and, for that matter, any medical writing, cannot exist or will not flourish without intense determination to learn and practice this craft.[1-3] The following are some of the main causes of bad surgical or medical writing.[1-3]

No Training and Little Writing Practice

Teachers of scientific writing would agree that the ability to write clear, concise, well-organized prose is a skill that can be learned. And therefore it can be taught. The problem resides, then, on lack of training and continuous practice.

Unless we change college curricula that allow science majors to write and practice this skill, our future surgical or medical practitioners will not become effective writers. It is disconcerting that "many of our best and brightest students are getting neither training nor practice in writing".[4] This area is also largely ignored by medical school and residency review committees.

Even with a concerted effort to teach writing, we should expect this skill to mature gradually. It takes many years beyond the completion of residency, well into high practice years, before the surgeon begins to harvest the fruits of years of intense writing. It has taken me (LHTP) more than twenty years since the completion of my surgical residency to reap the good deeds of committed writing.[15-18] Without starting this process earlier, it will be increasingly difficult to compete with nonphysician scientists in the scientific world.

No Extra Time for Writing

In general, biological sciences and medical students have a challenging heritage difficult to overcome. That is the chronic complaint about the lack of extra time for learning how to write and practice this significant skill. Finding the time is the student's and mentor's responsibility. Both should encourage themselves in cultivating the passion for writing. Under these conditions, it is not hard to find the time.

Lack of Instructors

To find well-accomplished writing surgical or medical instructors can be challenging. Few surgeons or physicians have completed writing classes that qualify them as trained writers. Most of us have learned by trial and error,[4] sometimes handicapped by a foreign primary language. In this situation, commitment and attention to detail are of paramount importance.

Unattended Competition

Competition is part of the writing process. Whether it is in the form of an article or a book chapter, the surgeon or medical writer needs to present the best possible works. Errors in grammar, punctuation, and syntax will detract potential reviewers from the significance of the presented work. Poor writing will diminish the opportunities of reaching publication. Attention to all writing rules is especially important and often a matter of acceptance or rejection.[5] A clear explanation of the topic being discussed will complete the paper's presentation and provide a positive backdrop for the evaluation of the work.

The Recompense of Writing

Writing has many direct and indirect positive effects on the surgeon's career. The surgeon who writes scientific works has the ability to project his/her message to the rest of the profession and the public in general. Writing distinguished scientific works also creates an aura of knowledge that permeates around the conveyor of news, the author of surgical science.

Indirect beneficial effects of writing include greater understanding of the subject matter, enhanced reputation within the topics published, improved case referral and frequent invitations for lectures at academic medical centers and national congresses.

High-Quality Medical Writing

High-quality surgical or medical writing is clear, well-defined, and possesses a worthwhile message that is easily identifiable. The following are some of the qualities of good medical writing according to various experts in the field.[1-18]

Thinking Like a Reader

The reader is the ultimate judge in assessing the quality of scientific surgical writing or general public writing. By thinking like a reader, the writer can prevent errors always present in the work of over-confident writers, namely leaving unexplained gaps in introducing concepts and providing inadequate background information.

The reader is an effective reviewer and, therefore, the surgical scientific writer will benefit a great deal from the reader's input. The reader offers advice from a perspective not frequently received from colleagues more intimately involved in the work. Following steps to maximize efficiency and clarity are essential and are aided by practice and constructive critiques.

Be Purposeful

Purpose in writing is mandatory if one is to excel in surgical, medical, or general writing. To have purpose is to have a well-defined idea that advances the understanding for others. To have purpose is to offer a plan that will clearly explain the aims and overall development of the topic in question. To have purpose or to be purposeful is to engage yourself in elements of good writing, in elements of positive thinking, and/or in elements of direction, keeping the ultimate goal of excellent writing in mind.[2,8] Every sentence or paragragh should advance toward the objectives initially outlined at the outset of the writing.

The Need for Clarity

There is no substitute for clarity of message, clarity of delivery, and clarity of expression. Clarity is the essential ingredient of any superior writing.[7-11] Clarity involves knowing what information is relevant or irrelevant and aligning the relevant information into a logical and concise presentation.

Surgical programs and medical schools need to establish writing lectures which emphasize how to write with clarity and elegance. It would be the emergence of these lectures that would encourage the development of the surgical scientific writing leaders of tomorrow.

Improve Your Writing

Practice All the Time

Like surgical skills, writing skills must be nurtured and honed through practice. A dedicated surgical or medical scientific writer remains continuously practicing his/her skills. In the same way that body builders regularly attend their gym practices, surgeon writers need to persistently write in their area of expertise to maintain their skill.

For medical students, residents, and young colleagues, we need to establish new curricula that would be amenable to improving their writing abilities. Other writing experts have recommended that "a good practice is to photocopy a journal article taking out the abstract and discussion sections and having the students/residents read the article and write an abstract or the discussion for it".[1-4]

Get an Editor to Help You

Most centers from which many papers get published (e.g., Johns Hopkins, Mayo Clinic, Cleveland Clinic, University of California at San Francisco, etc.) have something in common: medical editors. Medical editors do more than just "fix it up". When reading a surgical or medical manuscript, good medical editors look for meaning, organization, and consistency. In addition, they are reference checkers, and software critics. Before they send a manuscript to a journal, good editors should know about: (a) journal editors; (b) possible reviewers; (c) journal specifications for manuscripts; (d) intended readers; and (e) recent articles on similar topics. In other words, medical editors have to know about the subject that is being published, aside from the real process of editing that involves grammar, punctuation, syntax, and others.[10]

Finally, Publish Your Own Work

Acceptance of scientific surgical or medical manuscripts is not an easy task. It does require a great deal of work. First is the integration of meaningful results; second is writing well these results; third is reviewing several times the presentation of the results; and fourth is selecting an appropriate journal for submission. It is not unusual that certain surgical and medical journals reject 60-70% of the submitted papers.[8,10,12-14]

General agreement among editors remains that "journals don't reject good articles because they have many of them. Good articles describing quality studies are accepted even if they cannot be published immediately. The three most common reasons that journals reject articles are: (a) manuscript is inappropriate; (b) manuscript describes poorly designed or conducted studies; (c) manuscripts are poorly written".[8,10,12-14]

Editors agree as well that inappropriate manuscripts mean that many writers are not doing their homework in reviewing the "Instructions for Authors" page. In some journals it's in every issue; in others it's in the January and July issues. Poorly designed studies are those that have insufficient information, inadequate samples, biased samples, confounding factors, vague endpoints, straying from the hypothesis, and poor control of numbers. A good practice is to have a colleague or a physician in another department review the manuscript before sending it off. Many editors and reviewers have received manuscripts so filled with errors that they were not possible to evaluate. The more eyes that see the manuscript before it goes out, the less chances it will have of being rejected. One particular way to pick up errors and inconsistencies is to read the manuscript aloud. Another way is to let the manuscript rest for a couple of weeks, then reading it again.[1-3,5,8,9,13,14]

In Conclusion

Perhaps the most important reason for becoming a good surgical scientific writer is to be able to communicate science in clinical or basic research at the highest level. Surgeons or other medical writers can better express their work if they have learned the fundamentals and practice the art of good writing. Teaching younger colleagues, residents, and students how to become good writers will effectively elevate the field of the surgical writing.

The advice for those who are beginning their surgical career is to master the skills of the discipline and avoid the pitfalls of not thinking like a reader, not having purpose, not refining the clarity in their writings, and not practicing all the time. One could frankly say, be a good surgical scientific writer and your message will be better appreciated by your intended audience.

References

1. Booth V. Communicating in Science: Writing a Scientific Paper and Speaking at Scientific Meetings. 2nd ed. Cambridge, England: Cambridge University Press, 1983.
2. Publication Manual of the American Physiological Association. 3rd ed. Washington, DC: American Physiological Association, 1983.
3. Zeiger M. Essentials of Writing Biomedical Research Papers. 1st ed. New York, NY: McGraw-Hill, Health Professional Division, 1991.
4. Bjork R. Writing courses in American medical schools. Journal of Medical Education 1983; 58(2):112-6.
5. Huth EJ. How to Write and Publish Papers in the Medical Sciences. 2nd ed. Baltimore: Williams and Wilkins, 1994.
6. Zinsser W. On Writing Well. 1st ed. New York: Harper and Row. 1980.
7. Strunk Jr W, White EB. The Elements of Style. 3rd ed. New York: The Macmillan Company, 1979.
8. Anaya-Prado R, González Ojeda A, Arenas Márquez H. Obtener la Revista Ideal. Un Equilibrio Difícil. Cir Gen 2001; 23(4):304-307.
9. Iverson C. American Medical Association Manual of Style. 8th ed. Baltimore: Williams and Wilkins, 1988.
10. CBE Style Manual Committee. Council of Biology Editors Style Manual: A Guide for Authors, Editors, and Publishers in the Biological Sciences. 5th ed. Bethesda: Council of Biology Editors, 1983.
11. Williams JM. Style: Ten Lessons in Clarity and Grace. 4th ed. New York: HarperCollins, 1994.
12. International Committee of Medical Journal Editors. Uniform Requirements for manuscripts submitted to biomedical journals. Annals of Internal Medicine 1982; 96:766-71.
13. Booth WC, Colomb GG, Williams JM. The Craft of Research. Chicago, IL, USA: The University of Chicago Press, Chicago and London, 1995.
14. James DS, Spiro H. Writing and Speaking for Excellence: A Guide for Physicians. Sudbury, Massachusetts: Jones and Bartlett Publishers, 1996.
15. Toledo-Pereyra LH. Compassion. J Invest Surg 2005; 18:157-60.
16. Toledo-Pereyra LH, Samuel D. Gross: The Nestor of American Surgery. J Invest Surg 2006; 19:141-5.
17. Toledo-Pereyra LH. Alfred Blalock: Surgeon, Educator, and Pioneer in Shock and Cardiac Research. J Invest Surg 2005; 18:161-5.
18. Toledo A. The Medical Legacy of Benjamin Rush. J Invest Surg 2004; 17:61-3.

Section II.
Virtues of Man

CHAPTER 7

Knowledge

Luis H. Toledo-Pereyra

The practice of surgery should be dependent on the use of well-established research methods and recognized proven techniques, preferably derived from randomized clinical studies or systematic analyses. The practice of surgery should be based on well-proven knowledge.

Surgery, like many other medical disciplines, has evolved by building personal experience upon personal experience. The knowledge was anecdotal and established through individual cases. Observation represented the most valuable ally under these conditions, and new advances were incorporated with a clear vision of modifying the pathophysiology encountered. Examples of this evolving system included most of the surgical procedures we know today, such as hernia repair, appendectomy, small bowel obstruction treatment, cancer surgery, organ transplantation, and cardiac surgery. These techniques offered extraordinary results and did not necessarily come from the system now called evidence-based surgery. How did this happen? Is it possible that evidence-based surgery is just a new name for the same system we have always utilized? Can we continue practicing surgery as we have always done with good results and not mention evidence-based surgery? Let me provide an answer for you.

For five years (1999-2004), while teaching the 18-hour annual course called "Critical Analysis/Analytical Medicine" (based entirely on the teachings of evidence-based medicine as proposed by Sackett and his group[1]) at Michigan State University College of Human Medicine in Kalamazoo, I initially thought that evidence-based medicine was not a new system. Indeed, I believed it to be the same system utilized since the ancient times of Hippocrates and the Greek natural school of medicine, continued by the great physicians Galen, Avicenna, Harvey, and many others into our present era. I could not believe, much less understand, how we had been practicing medicine and surgery for thousands of years-before evidence-based medicine

Reminiscences on Surgery, History and Humanities,
edited by Luis H. Toledo-Pereyra. ©2007 Landes Bioscience.

debuted in 1992[1-3]—incorporating progress, innovative knowledge, and new ways of treatment, supposedly without using evidence-based medicine and surgery. To me, this created an incongruency that was difficult to accept!

During my surgical residency at the University of Minnesota (1970-1976), I felt that science and research were at the forefront of our surgical activities, though without the name of evidence-based surgery. Other institutions of higher learning also had similar approaches to surgery without using the specific description of evidence-based. Years after the term was coined (1992), surgeons continued to make incredible discoveries based on a traditional understanding of science, research, and clinical experience. Where was evidence-based surgery then? And is it needed now?

It was not until I began teaching "Critical Analysis/Analytical Medicine" again in 2005, that I started to understand the enormous value that evidence-based medicine or surgery supplies. I guess it took me five years of study and teaching to better understand this approach to education. This special system's roots finally revealed its potential to me. I am a complete believer now and let me tell you why. Evidence-based surgery brought with it the application of systematic research to the traditional practice of surgery. It elevated the means by which a surgeon, an evidence-based surgeon, practices surgery and interprets and evaluates the literature to apply it ideally, that is with the understanding that statistics and study design enter into the final assessment of the patient or study results. According to evidence-based principles, the surgeon should always ask three questions: Are the results valid? What are the results? and Are the results applicable to my patient?[1-12]

If David Sackett, distinguished clinician working until recently at McMaster University in Hamilton, Ontario, is considered the father of evidence-based medicine, certainly Archie Cochrane (1909-1988), noted British epidemiologist, should be considered the grandfather of this exciting medical movement. Cochrane was a dedicated proponent of randomized clinical trials, because he believed in their ability to provide the best clinical evidence when no one else was much interested. In 1972, he published his monumental work, *Effectiveness and Efficiency: Random Reflections on Health Services.*[13,14] In it he strongly supported the utilization of randomized trials to improve results and to test interventions. Even though the first randomized clinical trials came out in 1948 testing the role of streptomycin in treating tuberculosis, clinical epidemiology using randomized trials did not take hold until after Cochrane's book was published in 1972.[13-16] In recognition of Cochrane's pioneering contributions to evidence-based medicine, Oxford University opened the first Cochrane Center in 1992, and a year later began the Cochrane Collaboration with other centers around the world.[17-19]

Cochrane set the stage for Sackett and his group, who helped the surgical disciplines to accept evidence-based medicine as the basis for treatment and practice. Surgeons, encouraged by Cochrane's and Sackett's principles, addressed many key questions: How can we apply evidence-based medicine in a practical manner? What process allows us to implement evidence-based medicine effectively? What is evidence-based surgery? How can we qualify the evidence? How can we categorize the recommendations?

A definition of evidence-based surgery is extremely important, and it has already been presented at the beginning of this paper. Let us add the definition given by the Centre for Evidence-Based Medicine: "Evidence-based medicine is the conscientious, explicit, and judicious use of current best evidence in making decisions about the care of individual patients".[20,21] The surgical ideal is making the best surgical decisions possible based on the gathering of the best evidence currently available. In order to reach this ideal, surgeons must access the literature that provides them with relevant information.

- First, identify the topic.
- Second, go to the Cochrane database and systematic reviews.
- Third, proceed to Pubmed/Medline.
- Fourth, advance to e-Medicine and *Clinical Evidence.*

These steps facilitate an exhaustive review that permits you to ascertain, without a doubt, the real and current status of the medical literature. The most updated and extensive database is Pubmed/Medline. The most carefully evaluated database for evidence-based medicine is the Cochrane database.

In order to characterize the type of evidence, we need to stratify the evidence qualitatively. In this regard, the U.S. Preventative Services Task Force has delineated five levels of evidence by defining the studies from which the clinical data emanated.[10] Level I is the highest evidence obtained from at least one randomized controlled trial. Level II-1 is the next in importance, with evidence gathered from a well-designed controlled trial without randomization; level II-2 is evidence obtained from well-designed cohort or case-control studies, preferably from several centers; level II-3 is evidence accumulated from multiple time series with or without intervention. Finally, level

III, very anecdotal, is based in opinions of respected authorities, clinical experience, or reports of expert committees. This grading system, as good as it is, is not perfect since the relevance of measurements, the results of the study, and the baseline risk of the effect are not always included.[10]

Another concern is the categorization of recommendations based on the level of evidence. The U.S. Preventative Service Task Force has defined three levels: level A, where the recommendations are based on good and consistent scientific evidence; level B, where the recommendations are based on limited or inconsistent scientific evidence; and level C, where the recommendations are based primarily on consensus and expert opinion.[10] With this information, we can effectively categorize the level of evidence with the appropriate recommendations by strictly relying on the findings.

Robin McLeod presented an excellent review of evidence-based surgery from the surgeon's viewpoint in particular.[9] She discussed all elements of evidence-based medicine and surgery and reviewed issues of increasing importance for the practicing surgeon.[9] Issues of special concern for surgical trials were the standardization of the procedure and the surgeon's experience and ability to do the surgery. All aspects of perioperative and postoperative care should be standardized as well. One more point would be the performance of blind studies or sham operations which would be ethically difficult to justify in surgical trials today. Surgeons, then, should be aware of all of the various details in the practice of evidence-based surgery.[22-29]

Before concluding, I should mention some statistics pertaining to evidence-based medicine. Around 9,000 randomized controlled trials and over two million articles in 10,000 medical journals are published every year.[30] This reflects the incredible and difficult task facing practicing physicians and surgeons who wish to keep up with this continuous and ever-increasing body of evidence. For example, an internist would have to read approximately 19 articles each day to stay current.[30] General surgeons would probably be required to review a similar number of papers daily to remain well informed. The impossibility of carrying out this awesome task requires that surgical decisions today be based frequently on personal expertise, the weakest form of evidence. In spite of the constraints and limitations of today's busy clinical work, future surgeons must learn to efficiently practice evidence-based surgery so that they can use the strongest form of evidence, the evidence based on scientific findings and well-established clinical trials.

How can evidence-based surgery be taught at the highest level of efficiency without seriously curtailing the surgeon's current activities? I believe it can be done by incorporating the principles of evidence-based medicine into surgeons' daily rounds, operating room functions, postoperative treatment, and office practice. Only by integrating evidence-based principles into practice can we make our task simpler and the results worthwhile. When everyone practices evidence-based surgery, our ultimate objective of better patient care will be fully accomplished. We are looking forward to this day in the not too distant future. We are searching for the use of knowledge at all times!

References

1. Sackett DL, Richardson WS, Rosenberg W et al. Evidence-Based Medicine: How To Practice and Teach EBM. Edinburgh: Churchill Livingstone, 1998.
2. Sackett DL, Richardson WS, Rosenberg W et al. How to Practice and Teach EBM. New York: Churchill Livingstone, 1997.
3. Sackett DL, Rosenberg WMC, Muir Gray JA et al. Evidence based medicine: What it is and what it isn't. BMJ 1996; 312:71-72, (Available at: http://bmj.bmjjournals.com/cgi/content/full/312/7023/71?eaf%2523R2. Accessed August 1, 2005).
4. Straus SE, Richardson WS, Glasziou P et al. Evidence-Based Medicine: How to Practice and Teach EBM. Edinburgh: Elsevier Churchill Livingstone, 2005.
5. Muir Gray JA. Evidence-Based Healthcare: How To Make Health Policy and Management Decisions. Edinburgh: Churchill Livingstone, 2001.
6. Siwek J, Gourlay ML, Slawson DC et al. How to write an evidence-based clinical review article. Am Fam Physician 2002; 65(2), (Available at: http://www.aafp.org/afp/2002115/251.html. Accessed August 1, 2005).
7. Olkin I. Statistical and theoretical considerations in meta-analysis. J Clin Epidemiol 1995; 48:133-146.
8. Ellis J, Mulligan I, Rowe J et al. Inpatient general medicine is evidence based. Lancet 1995; 346:407-110.
9. McLeod RS. Evidence-based surgery. In: Norton JA et al, eds. Surgery: Basic Science and Clinical Evidence. New York: Springer-Verlag, 2001.
10. Evidence-Based Medicine, (Available at: http://en.wikipedia.org/wiki/Evidence-based_medicine. Accessed August 1, 2005).
11. Barton S. Using clinical evidence. BMJ 2001; 322:503-504, (Available at: http://bmj.bmjjournals.com/cgi/content/full/322/7285/503. Accessed August 1, 2005).
12. Urschel JD, Goldsmith CH, Tandan VR et al. Users' guide to evidence-based surgery: How to use an article evaluating surgical interventions. Can J Surg 2001; 44:95-100.

7

13. Culp K. History: Archie Cochrane, (Available at: http://www.smj.org.uk/0802/cochrane.htm. Accessed August 3, 2005).
14. Hill GB. Archie Cochrane and his legacy: An internal challenge to physicians' autonomy? J Clin Epidemiol 2000; 53(12):1189-92.
15. Alvarez-Dardet C, Ruiz MT. Thomas McKeown and Archibald Cochrane: A journey through the diffusion of their ideas. BMJ 1993; 306(6887):1252-4.
16. Archie Cochrane - Publications, (Available at: http://www.cardiff.ac.uk/schoolsanddivisions/divisions/insrv/libraryservices/research/cochrane. Accessed August 3, 2005).
17. Alvarao Nagib A. The Cochrane collaboration in the third world. Sao Paulo Med J 1998; 116(3), (Available at: http://www.scielo.br/scielo.php?script=sci_arttext&pid=S1516-31801998000300001. Accessed August 3, 2005).
18. Archie Cochrane Archive—Cardiff University, (Available at: http://www.cardiff.ac.uk/schoolsanddivisions/divisions/insrv/libraryservices/research/cochrane. Accessed August 3, 2005).
19. The name behind the 'Cochrane' Collaboration, (Available at: http://www.cochrane.org/docs/archieco.htm. Accessed August 2, 2005).
20. Scalise D. Evidence-based medicine. Hospitals and Health Networks 2004, (Available at: http://www.hhnmag.com/hhnmag/hospitalconnect/search/articl.jsp?dcrpath=HHNMAG. Accessed August 1, 2005).
21. Center for Health Evidence. Evidence-Based Medicine: A New Approach to Teaching and the Practice of Medicine. (Available at: http://www.cche.net/usersguides/ebm.asp. Accessed August 15, 2005).
22. Wente MN, Seiler CM, Uhl W et al. Perspectives of evidence-based surgery. Dig Surg 2003; 20(4):263-269.
23. Black N. Evidence-based surgery: A passing fad? World J Surg 1999; 23(8):789-793.
24. Saurland S, Lefering R, Neugebauer EAM. The pros and cons of evidence-based surgery. Langenbeck's Arch Surg 1999; 384(5):423-431.
25. Stirrat GM. Ethics and evidence based surgery. J Med Ethics 2004; 30:160-165.
26. Gentileschi P, Kini S, Catarci M et al. Evidence-based medicine: Open and laparoscopic bariatric surgery. Surg Endosc 2002; 16(5):736-744.
27. Baraldini V, Spitz L, Pierro A. Evidence-based operations in pediatric surgery. Pediatr Surg Int 1998; 13:331-335.
28. Searching for the Best Evidence in Clinical Journals, (Available at: http://www.cebm.net/searching.asp. Accessed August 1, 2005).
29. Gordon T, Cameron J, eds. Evidence-Based Surgery. B.C. Decker, 2000.
30. Antes G, Falck-Ytter Y, Trautmann N. Evidence-Based Medicine and the Cochrane Collaboration. (Available at: www.goinginternational.org/english/pdf/archive/Antes_ua_e.pdf. Accessed August 15, 2005).

CHAPTER 8

Compassion

Luis H. Toledo-Pereyra

> *Compassion is the quality or virtue of an individual to feel*
> *for someone else a strong sense of consideration for his/her*
> *despairing situation.*

The two-year-old girl was resting peacefully in the critical care unit of a Southern, well-staffed, medium-sized surgical specialty hospital. She had received multiple blood transfusions after severe splenic traumatic injury that required surgical intervention with spleen removal due to uncontrollable hemorrhage. Physically, everything looked fine up to this point. However, the surgeon, Phil Castellar, a middle-aged, abrupt, and unmannered professional, had previously indicated to father and mother that Rosie Mercado, the patient, would not require any organ extraction and that the surgery was for exploratory purposes only. He had gone on to say that the spleen should not be removed under any circumstances, especially in young children, because the procedure could make the patient more susceptible to infections. Mr. Mercado and his wife, Annie, were well aware of the state of affairs. They even remembered the doctor's exact words: "Do not let anyone touch the spleen. It will be tragic!"

When he finished this case, Brutus Phil, as he was sarcastically referred to by his detractors, directed himself to the waiting room and spent less than two minutes with the family, to whom he gave this final response: "The spleen is out and she has to live without it." He then abruptly left the room with virtually no opportunity for discussion and even less consideration for the worried parents. The anxious father and mother attempted to reach the surgical specialist, but he would not return their calls. Nurses in the unit attempted to pacify the irritated family as they eagerly awaited an answer from the operating surgeon, but no one could reach him. He could not be found!

Reminiscences on Surgery, History and Humanities,
edited by Luis H. Toledo-Pereyra. ©2007 Landes Bioscience.

Late the following morning, Dr. Castellar appeared before the caring parents with an unwarranted feisty attitude. Immediately upon his arrival, he indicated, "I do not like to be called for matters that are not important. You do not have the right to disturb me for emotional reasons. Now, what was the problem?" Mr. and Mrs. Mercado could not believe this unexpected and inappropriate response. They composed themselves and said, "Doctor, we were trying to talk to you about the situation with our little Rosie. You left without giving us all the details of her case, and we were very concerned about her condition." The experienced trauma and general surgeon did not know how to respond since he had just seen Rosie, and she looked amazingly well. "She is fine," he said in a tone of voice that clearly left no room for discussion. "But what about the spleen?" Mr. Mercado added. "You mentioned that no one should remove the spleen because it would be tragic." Even though the doctor recognized these as his exact words, he could not accept the burden of inappropriate behavior. "Who cares about the spleen now? She is alive, isn't she?" the frustrated surgeon responded. "With all my respect, doctor, our concern is with our little girl and what will happen to her in the future," said the quiet Annie Mercado, attempting to hold the surgeon's attention and obtain a satisfactory answer. Dr. Castellar could no longer contain himself and immediately stood up. Irritated for no reason with the whole event, he exclaimed, "We are not understanding each other. We are wasting our time discussing this matter. You need to talk to the pediatrician for future visits and follow-up." He then stomped off and demanded the nurse supervisor reach the pediatrician on call for follow-up and discharge.

Everyone in the nursing station immediately recognized the compassionless behavior of the knowledgeable and skilled trauma specialist. It was evident in the minds of the hospital administrators and physician educators that something had to be done with Dr. Castellar. Meanwhile, the Mercados could not comprehend the attitude of the famed surgeon or his failure to understand their concerns.

In as much as this case was not entirely real, the problem of lack of empathy and compassion within the medical community can at times be extraordinarily clear. The need for compassion in this realm is too important to let pass without formal acknowledgement and clear recognition by medical professionals (physicians, surgeons, and administrative officials). But beyond acknowledgement and recognition, what can be done to prepare medical professionals to appropriately deal with the despair of the sick and dying?

The question and well-defined challenge is how to educate students, surgical residents, and faculty regarding the exposure and specific development of professionalism and, in this particular circumstance, compassion in surgery.[1-14] This is no easy task, since many generations of surgeons before us have frequently approached the practice of surgery without emphasizing compassion as an important element of the healing surgical art. Surgeons past and present alike, as part of their training and their evolutionary process, as part of their mystique and the making of the professional, have not usually utilized compassion as one of the main components of the surgical career. Surgeons worry about learning the scientific basis of surgery, as well as mastering the surgical techniques needed to treat diseases surgically. And this is good, expected, and highly desirable. But in general, surgeons have not concentrated on exalting the value of compassion in their training or their surgical practice. This needs to change!

Surgeons need to find, exalt, and exude compassion in their professional practice. Surgeons should undergo theoretical and practical training not only in the science and techniques of surgery but also in the practice of compassion. Surgeons should enhance their ability to understand the suffering of others as well as the desire to do something about it.[3] This is the real test on the value of compassion. Surgeons need to become deeply aware of the strength of sympathy in relating to those who suffer.[3] Surgeons need to realize their own sorrow when faced with the despair of others.[2] Surgeons need to be agents of compassion as well as agents of cure.

History has not always demonstrated a positive influence on the use of compassion in surgery.[8] In fact, everything goes back to the management of pain, which has always been considered an acceptable companion to surgery. Even in the early nineteenth century, the great American physician Benjamin Rush (1746-1813) recommended heroic doses of painful remedies based on the belief that pain could cure.[8] In this atmosphere, there was no room for compassion. In the first century AD, Aulus Cornelius Celsus (3-64) defined the character of the surgeon in the following way: "Now a surgeon should be...filled with pity, so that he wishes to cure his patient, yet is not moved by his cries, to go too fast, or cut less than is necessary; but he does everything just as if the cries of pain cause him no emotion".[8] This attitude pervaded surgery for a long time.

Martin Pernick, in his well-researched book, *A Calculus of Suffering*,[8] clearly outlined the role of the surgeon in regards to pain: "For many early-nineteenth-century surgical students, learning to inflict pain according to these dicta of Celsus constituted the single hardest part of their professional training. Benjamin Rush's student Philip Syng Physick (1768-1837), the first American to gain prominence as a full-time surgeon, became so sick at his initial amputation that he had to be carried from the room in midoperation." A British doctor recalled one of his earliest surgical experiences:[8] "As the operation, which was necessarily a lengthy and slow one, proceeded, her cries became more and more terrible; first one and then another student fainted, and ultimately all but a determined few had left the theatre unable to stand the distressing scene." With this kind of information coming from the annals of surgical history, compassion did not yet have a proper place.

The introduction of ether anesthesia in 1846 began to change the attitudes and minds of practicing physicians and surgeons. James Y. Simpson (1811-1870), the Edinburgh physician and the discoverer of chloroform, in 1847 indicated that "the proud mission of the physician was distinctly two-fold, namely to alleviate human suffering as well as preserve human life".[8] Slow but progressive acceptance of pain control was beginning to take root within common surgical practice. Compassion was part of this movement. Years later, the famed literary figure T. S. Elliot (1888-1965) introduced the sentimental romanticism of his art into the expression of compassion in the healer's profession.[8] He composed the following ode to compassion and life:

SHARP COMPASSION:
THE SUPREMACY OF LIFE

The wounded surgeon plies his steel
That questions the distempered part;
Beneath the bleeding hands we feel
The sharp compassion of the healer's art.

However, quite different from a literary understanding of its necessity in medical practice is the practical aspect of teaching and practicing compassion. How can we establish an educational program on compassion that will be helpful in the training of surgeons? How can we develop standardized ways to teach young colleagues the importance of compassion in the routine practice of surgery? Undoubtedly, virtues of this magnitude are not simple to teach. In this case, I would call compassion—from the Latin *compati* (to suffer with another)—a virtue and not a skill, since virtue is "the readiness or disposition of man's powers directing them to some goodness of act".[5] Virtue is doing "what is right and not what is wrong".[4] And in these terms, then, compassion applies directly to being a virtue of the highest order. It is the virtue of supporting somebody in need, somebody in despair, the virtue of sympathizing and assisting others in distress.

Compassion can be taught in the classroom, on the floor, in the clinic, in the operating room, and throughout the entire surgical and clinical enterprise. Compassion can be taught by exemplifying and demonstrating the qualities associated with its practice. Compassion should be, and needs to be, part of the surgeon's daily clinical experience. Compassion should be an intricate part of the surgeon's world.

In regards to a specific teaching plan, the faculty should first be taught to identify three phases of compassion: recognition, acceptance, and participation. These phases are to be carefully considered and exemplified in practice for everyone to understand and later on to teach. In the *Medicine of Compassion*, Karen and Simon Fox analyzed the core skills needed to understand and review the compassion response.[14] Their interesting book and well-produced video are extremely helpful aids in the commitment to the teaching of this virtue. In this curriculum, they introduced four elements of compassion: acknowledgement, affection, acceptance, and attention. I prefer the initial three phases, even though either way is appropriate. Once the faculty has been assimilated into this process, residents and students should follow the same path. Specific examples from the floor, operating room, and classroom will compliment the knowledge and acceptance of compassion. Finally, weekly presentations and examinations will complete the full circle of teaching compassion.

A good example of the value of compassion in surgical practice was extremely well represented in a recent movie, *The Doctor*. In it, Oscar winning actor William Hurt characterized Jack McKee, the thoracic and cardiovascular surgeon who was a superb technician but a professional without compassion, a professional who believed that the surgeon's job was "to cut." He was a professional who relied more on the prowess of the knife than on the expression of the heart, a professional who had no sensitivity toward the care of the human being. As Dr. McKee put it, "I'd rather cut straight and care less."

But one event transformed everything! Jack McKee was diagnosed with cancer of the larynx, and the doctor became the patient. His experience in the diagnosis and management of the disease was so real and devastating to his human self, that he realized the worth of understanding and knowing the real individual behind the patient, the importance of offering empathy for someone else's suffering, the significance of caring for the impaired soul. For the first time in his life, JackMcKee realized the value of compassion.

As Jack McKee the patient developed a close friendship with June, a patient with an incurable brain tumor, he could not understand her strength in tolerating the effects of her disease. On one occasion of sincere empathetic feeling, he asked her, "Do you pray, June? Is that what keeps you together?" June very bravely responded by saying, "I pray and meditate, I eat chocolate, I go dancing." What an incredible moment of soul opening and complete acceptance of her place on earth, what a moment of conversion for the unemotional doctor, what a moment of integration of life, hope, and suffering, what a special moment of truth, reality, and love. Even though June died a few days later, Jack McKee, the previously callous doctor, had already changed his life, had already accepted compassion as the main ingredient of caring for another human being, had already become a real surgeon! It is in this sense that this overwhelming story is extraordinarily unique for its impact on the practice of surgery today.

It is ironic to think that the advice doctors and surgeons once received of "not getting emotionally involved with our patients" is exactly the opposite of what we are so rightly being recommended to practice today. It is as if somebody is saying, "Be a good surgeon, be a good technician, and above all, be a kind and generous human being. Be a compassionate surgeon."

References

1. Dictionary of Everything. (Available at: http://www.dictionaryofeverything.com/. Accessed June 7, 2005).
2. Classical Authors Index. (Available at: http://selfknowledge.com/18785.htm. Accessed June 7, 2005).
3. The Free Dictionary. (Available at: http://www.thefreedictionary.com/compassion. Accessed June 7, 2005).
4. WordNet Search. (Available at: http://www.cogsci.princeton.edu/cgi-bin/webwn2.1. Accessed June 7, 2005).
5. Wikipedia. (Available at: http://en.wikipedia.org/wiki/Virtue. Accessed June 7, 2005).
6. Glosario de te´rminos filoso´ficos. (Available at: http://www.filosofia.net/materiales/rec/glosaen.htm. Accessed June 7, 2005).
7. Sobel D. In: Solen D, Cohen J, eds. Introduction in The Best American Science Writing 2004. New York: Harper Collins Publishers, 2004.

8. Pernick MS. A Calculus of Suffering: Pain, Professionalism, and Anesthesia in Nineteenth-Century America. New York: Columbia University Press, 1985.

9. Othersen Jr HB. Ephraim McDowell: The Qualities of a Good Surgeon. Ann Surg 2004; 239:648-650.

10. Spiro HM, Curmem MGM, Peschel E et al. Empathy and the Practice of Medicine. New Haven: Yale University Press, 2003.

11. Almy TP, Colby KK, Zubkoff M et al. Health, Society and the Physician. Ann Intern Med 1992; 116:569-574.

12. Engel GL. Physician-scientists and scientific physicians: Resolving the humanism-science dichotomy. Am J Med 1987; 82:107-111.

13. Spiro H. What is empathy and can it be taught? Ann Intern Med 1992; 116:843-846.

14. Fox K, Fox S. The Medicine of Compassion: Video and Reader's Guide. Santa Barbara: Adventures in Caring Foundation, 2004.

CHAPTER 9

Respect

Luis H. Toledo-Pereyra

> *Respect is the cohesive element in the caring and understanding for someone that binds us all together. Respect is the great opportunity for enhancing the human being. Respect is an uplifting event of unique and expanding proportions.*

Respect is the engine that supports human behavior. Without respect civilizations would crumble. Individuals, therefore, require respect to maintain society's interactions. The surgical world is not that different from society. Surgeons work and live within the norms and principles of respect. Respect for their fellow surgeons, for the operating room personnel and environment, and, of course, for patients and hospital activity is essential.

Through the years, respect has not always been well practiced in the confines of the operating room and particularly with regard to the function and supervision of surgical residents. Let me be more specific. Surgeon teachers not infrequently demand efficiency and exactness of surgical residents in performing the operating act. I believe this is a good teaching practice. What is not good is to overuse the teaching capacity and exert unneeded demands from residents and fellows who are learning how to operate. What is not good is to categorize some residents as unfit even before working with them. What is not good is to create an atmosphere of tension during surgery that helps no one and adds one more burden of uneasiness for the learner. Let's then be conscious of the function and value of teaching and offer respect for residents and coworkers alike, during and after the operating act.

A young surgeon in training, Peter Svankel, was commiserating about his performance in room 14 of a respected Midwestern regional hospital, where he was helping Dr. Richard Versad, the well-known and noted academic surgeon who at the sunset of his career wanted to demonstrate his dogmatic and intransigent person-

Reminiscences on Surgery, History and Humanities,
edited by Luis H. Toledo-Pereyra. ©2007 Landes Bioscience.

ality. "Any movement I take, anything I will do, he will oppose or have a negative or derogatory comment about," Peter said to a fellow second-year surgical resident. The other resident, who had scrubbed with Versad before, clearly supported Peter's laments. "If it would be of some help, I can tell you that Versad takes the same attitude towards most of the residents, particularly when they are foreigners, and especially from third-world countries," he said. "How is that possible in the twenty-first century?" Peter could not comprehend how Versad would act that way and still be tolerated. Lack of respect and consideration permeated the whole enterprise, overruled human understanding, and remained evident in many surgical operating rooms around the country. Even though this is a fictional example, it closely represented the experience of many residents.

Just three decades ago, in the early 1970s, surgical residents, such as myself, regularly faced one, two, or more Versads in their academic training. Respect for surgical residents ranked low in good surgical programs, which prized surgical knowledge and practice. The surgical humanities and professionalism as a whole were low priorities, if they were priorities at all. Things have changed today, have they not?

So respect represents an essential condition for surgeon training. "Respect is the objective, unbiased consideration and regard for the rights, values, beliefs, and property of all people".[1] To accept and preach respect is our responsibility as human beings and as professionals directly involved in the nurturing and development of our respective careers. Having respect for someone means caring for this person, understanding his/her philosophy of life, being considerate in the interpretation and appreciation of his/her work. In essence, to be respectful is to understand who we are and how we react and function.[2-15]

Respect comes in many guises: respect of an individual, society, institution, or government. Respect of laws, family, school, or work conditions. Respect at work of fellow employees, superiors, and subordinates. Respect of minorities, the weak, the disabled, as well as of the strong and governing class. Respect should be for everyone and everything. One could easily say, practice respect and you will receive it back with praises. Respect can build bridges not dreamed of before. When we lose the perception and practice of respect in our lives, we lose our communication and closeness with others, whether colleagues, family, or coworkers. It does not matter, we have lost contact with our surrounding and supporting world!

Another type of respect, as important or more important than those previously mentioned, is the respect for ourselves. Self-respect is intrinsically related to opportunities for a successful and productive life. Self-respect is not a moral requirement, but it is a moral necessity. Self-respect is the most important moral duty according to the eminent German philosopher Immanuel Kant (1724-1804).[2] He explains, "Just as we have a moral duty to respect others as persons, so we have a moral duty to respect ourselves as persons, a duty that derives from our dignity as rational beings. This duty requires us to act always in an awareness of our dignity and so to act only in ways that are consistent with our status as an end in ourselves and to refrain from acting in ways that abase, degrade, defile, or disavow our rational nature. That is, we have a duty of recognition of self-respect".[2]

We need, then, to enhance our self-respect in order to fulfill our most essential moral necessity and, according to Kant's well-respected philosophical concepts, to fulfill our moral duty. Other philosophers[2] concur with Kant by reaffirming the importance of self-respect and its implications on social, political, and moral grounds. It is clearly apparent that individuals and organizations need self-respect to maximize their function and sense of worth in society.

The recognition of respect as an important quality for all people is well documented in the world of letters, philosophy, and science. Confucius (551-479 BC), a Chinese philosopher and reformer, said, "Respect yourself and others will respect you".[16] James Howell, a noted writer, conveyed, "Respect a man, he will do the more".[16] Cicero (106-43 BC), the extraordinary Roman thinker, advised, "He removes the greatest ornament of friendship who takes away from it respect".[16] John Herschel (1792-1871), prominent English mathematician and astronomer, indicated, "Self-respect is the cornerstone of all virtue".[16] In the same way, a Spanish proverb proclaims, "If you want to be respected, you must respect yourself".[16] Ralph Waldo Emerson (1803-1882), an eminent American philosopher, summarized very well our thinking on surgical training, "The secret of education lies in respecting your pupil".[16] Many other quotations celebrate the value of respect, but for today, we shall content ourselves with these.

Now, let us return to the surgical arena, where staff surgeons, surgical residents, surgical nurses, and the rest of the surgery personnel are freely integrated into a full and complete surgical team. Respect is elemental for any team that cannot function without everyone's participation. Respect is needed from each one of the team members. If one fails, the team will not show the congruency and efficiency so much desired for its effective function. Respect needs to be encouraged and be brought to the forefront of the surgical scene.

We do not have many well-documented examples in the annals of surgery that reflect the way respect was approached. However, patient respect was transparent in the writings of the father of medicine, Hippocrates (460-377 BC).[17] With Galen (130-200 AD), several centuries later, we do not encounter a similar approach, directly or indirectly, to that of the Hippocratic writings.[17,18]

During the Middle Ages, surgeons did not follow or express the path of respect for patients or other surgeons.[18] It was not until the greatest Renaissance surgeon, Ambroise Paré (1510-1585), that respect for patients was revived.[18] Respect for colleagues and disciples was not mentioned in a demonstrable manner. Many other surgeons of great esteem followed, and in spite of their significant surgical contributions, they were not necessarily oriented towards the humanistic side of surgery. Respect lagged even as surgery progressed.[19]

In the nineteenth century, few surgeons were evidently attracted by the teaching and practice of respect. One who did recommend complete respect for patients was Johan von Mikulicz-Radecki (1850-1905) from Breslau, at that time part of Germany but today in Poland. Unfortunately, he was "a ruthless master to his collaborators".[18] Sadly, then, no consideration of respect for associates or trainees was contemplated. Mikulicz had been a disciple of Theodor Billroth (1829-1894).[18] Did the indifference to respect become a trend, followed by surgeons in the treatment of their trainees, which persists to our time?

At the dawn of the modern age of surgery, William Halsted (1852-1922) of Hopkins fame, occupied a place of distinction in the surgical world.[19] His principles and recommendations were highly respected. How he respected students, residents, and colleagues is not completely recorded. However, from the prodigious pen of the New Orleans surgeon Rudolph Matas (1860-1957),[20] Professor Halsted emerges as a kind and courteous individual even though seriousness, calmness, and shyness were indelible marks of his personality. We see the noted Baltimore surgeon as offering respect for his surgical team, in and out of the operating room, a great quality for the pioneering specialist, something that should be emulated by future generations of surgeon masters.

In contemporary times, as recently as fifty years ago, while a fresh high school student in Mexico, I had the great fortune of participating in the operating room with Dr. Victor Manuel Romo Ruíz, a distinguished surgeon who was highly respected by all in town. He exuded competency, but most of all demonstrated a deep, caring respect for everyone attending the surgical act. He was the great example students were searching for, the professional we wanted to be associated with, the extraordinary human being we needed to keep in mind. During my medical schooling in Mexico City at the National Autonomous University of Mexico, I became impressed with another outstanding surgeon and teacher, Dr. Gilberto Lozano Saldívar. He would attend academic patients at Juarez Hospital every morning and teach medical students and residents the highlights of surgery. He was a caring and especially gifted human being. He highly respected all members of the surgical team, he appropriately praised to lift morale, and he made everyone feel an important part of the team. It was unfortunate that the earthquake of 1982 in Mexico City put an end to his caring and productive life.

In the 1970s, I was a surgical resident at a great American medical center in Minneapolis, the University of Minnesota. I learned a great number of very unique ideas and concepts about surgery. I was exposed to the best teaching and most advanced operative experience. But I realized as well that some mentors were more interested in showing respect to members of the surgical team than were others. Surgeons who lacked respect created an uncomfortable environment for everyone in the operating theatre. Under these conditions, residents became the most vulnerable group, followed thereafter by the scrub nurses and surgical technicians. How to teach surgery and remain respectful? How to operate well and not lose your temper? How to face adversity during surgery and react with equanimity? How to be a good surgeon and above all be a good and respectful human being? These are the genuine and fundamental thoughts that should guide today's surgeons. Most importantly, how can we teach respect to faculty, residents, and nurses alike within the operating room boundaries?

Certainly we can start by recognizing the presence or absence of respect, by documenting lapses, by praising those who practice respect, and by offering educational and tutorial support to those who do not. Several ideas come to mind, such as establishing seminars, developing randomized videotaping that clarifies acceptable and unacceptable behavior, instituting role playing between residents and practicing surgeons, and finally, incorporating surgical resident advocates, who will exercise fairness to all involved, into surgical residencies.

In the coming weeks, I challenge you to be aware of the various circumstances in which, by observing respect, you can make a difference in the field of surgery and in the many individuals who practice this noble profession. In closing, respect is the cohesive element in the caring and understanding for someone that binds us all together. Respect is a great opportunity for enhancing the human being. Respect is an uplifting event of unique and expanding proportions.

References

1. Respect. (Available at: http://en.wikipedia.org/wiki/Resepct. Accessed July 5, 2005).
2. Respect. Stanford Encyclopedia of Philosophy. (Available at: http://plato.stanford.edu/entries/respect. Accessed July 5, 2005).
3. Baron MW. Love and respect in the doctrine of virtue. The Southern Journal of Philosophy 1997; 36(Suppl):29-44.
4. Blum A. On respect. Philosophical Inquiry 1988; 10:58-63.
5. Cranor CF. Kant's respect-for-persons principle. International Studies in Philosophy 1980; 19-40.
6. DeMarco JP. Respect for persons: Some prerequisites. Philosophy in Context 1974; 3:33-37.
7. Ghosh-Dastidar K. Respect for persons and self-respect: Western and Indian. Journal of Indian Council of Philosophical Research 1987; 5:83-93.
8. Hudson SD. The nature of respect. Social Theory and Practice 1980; 6:69-90.
9. Sherman N. Concrete kantian respect. Social Philosophy and Policy 1998; 15:119-148.
10. Spelman EV. On Treating persons and persons. Ethics 1977; 88:150-161.
11. Wood AW. Kant's Ethical Thought. Cambridge: Cambridge University Press, 1999.
12. In: Dillon RS, ed. Dignity, Character, and Self-Respect. New York: Routledge, 1995.
13. The RESPECT Project. (Available at: http://www.respectproject.org/main/index.php. Accessed July 5, 2005).
14. The Six Pillars of Character-Respect-Popcorn Park. (Available at: http://www.goodcharacter.com/pp/respect.html. Accessed July 6, 2005).
15. The Foundation for a Better Life. (Available at: http://www.forbetterlife.org/values/index.asp?section=values&valueID=45&language=eng. Accessed July 5, 2005).
16. The Quotations Page-Your Source for Famous Quotes. (Available at: http://www.quotationspage.com/. Accessed September 9, 2005).
17. Hippocrates. Volume I. WHS Jones, transl. Cambridge: Harvard University Press, 1923.
18. Haeger K. The Illustrated History of Surgery. New York: Bell Publishing Company, 1990.

19. Toledo-Pereyra LH. Vignettes on Surgery, History and Humanities. Georgetown: Landes Bioscience, 2005.
20. Matas R. William Stewart Halsted, 1852-1922: An Appreciation. Johns Hopkins Hosp Bull 1925; 36:2.

9

CHAPTER 10

Integrity

Luis H. Toledo-Pereyra

> *Integrity is the dedicated commitment to an accepted conduct and form of living . . . is the honest response to professional attitudes and values.*

Since the dawn of civilization, integrity has represented the base on which respectable societies are built.[1-11] The society of surgeons is no exception. Surgeons require integrity to remain firm in the execution of the surgical act and in the care of the surgical patient. Integrity leads to respect, which, in turn, leads to understanding and agreement. Integrity is honesty with something else, the opportunity to reiterate your position under normal and difficult circumstances. What can we say about integrity that has not been said before? How can we emphasize integrity and maintain interest in its practices and evolution? Where can we reach for continuous support? These are just a few of the questions that appear in our search for integrity.

Now, let us concentrate on the surgical arena and evaluate the importance of integrity in the surgeon's life and career. Let us define the character of integrity and advance its effect. As the senior surgeon was getting ready for his daily rounds, the inexperienced first-year resident had an important question. "Dr. Mackey," he said, "do you think we should tell the patient everything that occurs in the operating room during the surgical act?"

Replied Dr. Mackey, "I believe you have the professional responsibility to mention to the patient all important and significant parts of the operation."

"What about discussing with the family the mistakes that occurred during the operative event?" asked the junior resident.

"I am of the belief, since the early times of my surgical training, that surgeons need to communicate to their patients any negative developments that take place during surgery," added Mackey. "This is an important issue that speaks to the integrity and moral fiber of the surgeon-in-charge."

Integrity is not a simple state but a complex condition reflecting an individual attitude towards a strict and accepted way of living. Surgeons with integrity are surgeons who keep their promises in spite of difficult situations. Surgeons with integrity are surgeons who speak the truth. Surgeons with integrity are surgeons who respect others and attend to their needs. Surgeons with integrity are surgeons who maintain an accountable and precise record. Surgeons with integrity are surgeons who follow their patients' needs.

Integrity is fully developed after dedicated teachings and persistent examples are given pertaining to the do's and don't's following the surgeon's world and interactions with patients, residents, and fellow surgeons. The history of surgery gives us excellent examples of the value of integrity in the surgeon's practice and academic teachings.

An unique and pertinent example of integrity is that of the first inoculator in America. Zabdiel Boylston (1680-1766), well-recognized pioneer American surgeon, had an extraordinary force of character and "strict adherence to the standard values and a way of conduct".[12] He remained loyal to his principles of defending the special virtues of inoculating patients with small-pox during the Boston epidemic of 1721. Several prominent physicians of the time such as Douglass and Dalhonde greatly opposed and ridiculed Boylston's practice.[12] Many in the city of Boston followed the stance of the medical attackers. Boylston, without hesitation, defended his just cause until the epidemic ravages temporarily terminated in the second half of 1722.[12] His conduct testified as to the force of character and integrity, without reproach, that marked his professional life and was demonstrated in an immaculate manner and at all times.

Another example of surgical integrity in America involves the great figure, Kentuckian surgeon master EphraimMcDowell (1771-1830).[13] Against all odds, in the backwoods of undeveloped Kentucky, he performed in 1809 the first successful abdominal operation in the world.[13-14] With dedicated interest and uncontested integrity, he explained to his doubtful patient, Jane Todd Crawford, the potential outcome of an operation such as the one she needed. With uncanny honesty, he indicated to her that no successful abdominal operation had been completed before and that her operation was an experiment that had never been accomplished with positive results.[13,14] In spite of these unsettling premonitions, Mrs. Crawford decided to proceed with the surgery with the hope that her surgeon was ready. McDowell and Crawford went through the surgical act with great strength and determination. The operation was a resounding success, the patient tolerated the procedure well in spite of the lack of anesthesia, her ovarian tumor was fully excised, and the surgeon removed it in one piece! Other successful similar operations were performed byMcDowell in subsequent years.[13,14]

In addition to the integrity exhibited by the courageous behavior of Boylston and McDowell, many other American surgeons demonstrated identical strong qualities.[15,16] Suffice it to say that in equal terms, Beaumont,Mott,Warren, Gross, and Halsted are just a few of a large number of nineteenth century American surgeons who demonstrated the enormous importance of integrity in their professional careers.

Dr. Mackey, our fictional senior surgeon, concerned with the proper demonstration of the value of integrity for surgical residents throughout the whole surgery program, began a careful study pertaining to the better understanding of integrity as a philosophical entity. In his exhaustive library search through evidence-based surgery,[11] he encountered the *Stanford Encyclopedia of Philosophy*[8] with the following fundamental concepts:

"Integrity is one of the most important and oft-cited of virtue terms. It is also perhaps the most puzzling. For example, while it is sometimes used virtually synonymously with 'moral,' we also at times distinguish acting morally from acting with integrity. Persons of integrity may in fact act immorally—though they would usually not know they are acting immorally. Thus one may acknowledge a person to have integrity even though that person may hold importantly mistaken moral views. When used as a virtue term, integrity refers to a quality of a person's character; however, there are other uses of the term. One may speak of the integrity of a wilderness region or an ecosystem, a computerized database, a defense system, a work of art, and so on. When it is applied to objects, integrity refers to the wholeness, intactness or purity of a thing—meanings that are sometimes carried over when it is applied to people. A wilderness region has integrity when it is not corrupted by development or by the side-effects of development, when it remains intact as wilderness. A database maintains its integrity as long as it remains uncorrupted by error; a defense system as long as it is not breached. A musical work might be said to have integrity when its musical structure has a certain completeness that is

*not intruded upon by uncoordinated, unrelated musical ideas;
that is, when it possesses a kind of musical wholeness, intactness and purity. Integrity is also attributed to various parts or
aspects of a person's life. We speak of attributes such as professional, intellectual and artistic integrity. However, the most
philosophically important sense of the term integrity relates to
general character. Philosophers have been particularly concerned
to understand what it is for a person to exhibit integrity
throughout life. Acting with integrity on some particularly
important occasion will, philosophically speaking, always be
explained in terms of broader features of a person's character
and life."*

Surgeon Mackey was extremely pleased with the surgical residents
exposure to integrity. However, the practicality of teaching this virtue was not completely clear. It was necessary then to develop a program that conveyed the best means of teaching the extraordinary
importance of integrity. This potential program would include theoretical principles as well as practical examples—as indicated before—
of the characterization of this significant professional and personal
trait. Mentors dealing with this issue should utilize all opportunities
available for referring to the presence of integrity at work and in their
lives. A dedicated and long-term task!

The University of Illinois Core Curriculum program[17] considers
honesty and integrity as two of seven very important qualities that
characterize the "true medical professional." In their words, "a physician should keep his or her word, regardless of personal cost, and
speak truthfully. Also physicians should avoid situations that place
their own interests in conflict with those of others".[17]

In closing, we need to remember that integrity is identified in
someone who remains true to his/her own principles, someone who
maintains a crystalline pure expression of character, as well as someone who exudes honesty in personal and professional behavior. Without integrity, the life of a surgeon, researcher, or academician is not
possible, because trust, belief, and understanding would be completely
eliminated and every action would be entirely meaningless.

References

1. Graham JL. Does integrity require moral goodness? Ratio 2001; 14:234-251.
2. Halfon M. Integrity: A Philosophical Inquiry. Philadelphia: Temple University Press, 1989.
3. Holley DM. Self-interest and integrity. Int Philos Q 2002; 42:5-22.
4. McFall L. Responsibilities of scientists and intellectuals. Routledge Encyclopaedia of Philosophy 1999.
5. Putman D. Integrity and moral development. J Value Inquiry 1996; 30:237-246.
6. Van Hooft S. Judgment, decision, and integrity. Philos Explorations 2001; 4:135-149.
7. Zagzebski L. Virtues of the Mind: An Inquiry into the Nature of Virtue and the Ethical Foundations of Knowledge. Cambridge: Cambridge University Press, 1996.
8. Integrity. Stanford Encyclopedia of Philosophy. (Available at: http://plato.stanford.edu/entries/integrity. Accessed November 9, 2005).
9. Toledo-Pereyra LH. Compassion. J Inv Surg 2005; 18:157-160.
10. Toledo-Pereyra LH. Respect. J Inv Surg 2005; 18:281-284.
11. Toledo-Pereyra LH. Evidence-based surgery. J Inv Surg 2005; 18:219-222.
12. Boylston, Dr. Zabdiel FRS. In: Thatcher J, ed. American Medical Biography, Volume 1. New York: DaCapo Press, 1967, (reprinted from first edition published in Boston in 1828).
13. Gray Sr L. The Life and Times of Ephraim McDowell. Louisville, Kentucky: VG Reed and Sons Printers, 1987.
14. Toledo-Pereyra LH. Ephraim McDowell: Father of Abdominal Surgery. J Inv Surg 2004; 17:237-238.
15. Webster II New Riverside University Dictionary. New York: Houghton Mifflin, 1984.
16. Toledo-Pereyra LH. Vignettes on Surgery, History and Humanities. Georgetown, Texas: Landes Bioscience, 2005.
17. The Board of Trustees of the University of Illinois. Online Graduate Medical Education Core Curriculum at the University of Illinois Chicago, 2003, (Available at: http://www.online.uillinois.edu/catalog/ProgramDetail.asp?ProgramID=293. Accessed December 5, 2005).

CHAPTER 11

Trust

Luis H. Toledo-Pereyra

To believe in the integrity, confidence, and reliance of another.

Imagine you were in the middle of the Pacific Ocean, there was not a soul you could depend on, and your life could only be saved by one person, a person you knew from college as a deceiving and untrustworthy human being. The reality and importance of this situation is evidently clear when considering trust in the life of a person, society, or the world, even if this story is fictional.

Now, imagine you were airborne in a commercial airliner, negotiating altitude at the eye of a heavy storm. And, image your life depended on the knowledge and ability of the plane's captain, who had suffered a mild stroke and was unable to perform optimally. The cocaptain, with less experience, immediately took over control of the faltering plane although with less predictable results. When considering this possibility, trust, again, represents the essence of continuing survival.

Imagine once more that your welfare relied on the dedication and participation of poorly prepared health professionals and that you were not sure you were receiving the best health care. Imagine you could not accept their care if your life depended on it. Your future, suddenly and dramatically, becomes threatened and purposeless. Trust is missing!

The previous examples reflect very clearly the far-reaching dimensions and overwhelming, encompassing effects of trust. Trust means to believe in the integrity and reliance of another. According to Webster's Dictionary, trust is "to have confidence in; to expect with assurance; to believe in someone in whom confidence is placed".[1] Confidence, at the same time, is to have "certainty in the trustworthiness of another".[1] Trust and confidence come from the same tree and represent the same family. They are basically the same word with a different wrapping. Confidence is "the assurance that someone will keep a secret".[1,2]

Reminiscences on Surgery, History and Humanities,
edited by Luis H. Toledo-Pereyra. ©2007 Landes Bioscience.

How is it then that surgeons utilize trust in the evolving and changing relationships of their profession? How is it that surgeons cannot proceed further without the enduring presence of trust? How is it that surgeons cannot practice their art or science under trustless circumstances? The response is so evidently clear that an explanation appears to be hardly required and mostly redundant. Surgery and trust are synonymous with good practice. Surgery and trust coexist at all times. Surgery and trust cannot exist apart from each other.

If trust is so uniquely important, how can we then teach it to new generations of practicing surgeons? If trust is so particularly inherent to good practice, how can we then advance its very much-needed presence? If trust is of such a paramount significance, how can we then characterize its value?

As we advance the quest for the teaching of values to the surgeons of tomorrow, I remain mindful of the complexity associated with this especially extraordinary endeavor. Many other virtues or human qualities, such as compassion,[3] respect,[4] integrity,[5] commitment, and now trust, continuously interact with each other. They intimately depend on the functions of the other. They cannot be orchestrated in practice without a consideration for one another. They all work together and practically at the same time. It is inconceivable to think that compassion could exist without integrity and that integrity could appear without commitment or trust. Compassion—respect—integrity—commitment and trust are, in the end, represented in the same congregation of uplifting and profound human values. We need them all to achieve a positive and worthwhile existence.

Let us now concentrate on the surgical arena where surgeons, residents, and assistants frequently enter the surgical theater to realize the most sacred act of healing. There trust constitutes an unique and particularly special quality that transcends the attributes of other functions, because without trust we cannot effectively advance the patient's surgical care. Trusting your surgical mentors, partners, and associates will permit the full integration of the operative act and will, at the same time, enhance the possibility of success.

Trust constitutes life for the surgeon. Without it, surgery is incomplete, surgery is not itself, surgery is highly deficient. To instill trust in the training and practice of the surgeon is vital. To understand and recognize anatomy, to understand and explain physiology, to understand and analyze pathology, and to understand and perform surgery, trust is fundamentally and practically essential. Trust is the missing link of the practicing surgeon. Trust is required for the evolution of surgery!

Trust in science, trust in personal surgical accomplishment, trust in human capability, and trust in patient care are all critical elements of the surgeon's world. It is imperative then to cultivate and continuously support endeavors directed at maintaining the highest standards of trust. When trust is reached, professional and personal qualities are clearly enhanced in an ideal manner. Consider trustworthiness as the most unique factor in the life of anyone,[2] in the life...of a surgeon.

From the times of Zabdiel Boylston (1679-1766), who first published accounts of elective surgical operations in the American British colonies,[6] to the extraordinary contributions of William S. Halsted (1852-1922),[7] trust has been at the forefront of the American surgical profession. Zabdiel Boylston exuded trust and demonstrated its significance as he dealt with the pioneering surgical cases of early eighteenth century America. Patients and families had trust in the activities and surgical ability of this Boston surgeon. When Boylston attended his patients, especially during the 1721 smallpox epidemic, he stood at the center of trust and confidence.

William Halsted, equally as accomplished as Zabdiel Boylston but living in modern times, clearly advanced the surgical sciences through his innovative capacity and his uniquely qualified personality, which stimulated and maintained trust at the cusp of his professional activity.[7] William Halsted constantly revealed his deep knowledge and unique intuition in the surgeon's profession, and by doing so, faculty, residents, and patients placed an incalculable amount of trust in all his surgical decisions and activities.

Other distinguished American surgeons of earlier times, such as John Jones (1729-1791), Philip Syng Physick (1768-1837), Ephraim McDowell (1771-1830), John Syng Dorsey (1783-1818), William Beaumont (1785-1853), Valentine Mott (1785-1865), and many more, brought with them, in various ways, a great deal of information, understanding, respect, compassion, trust, and integrity[7,8] to the American scene. Each one of these basically uplifting qualities was an intricate part of the life and work of these pioneer surgeons.

Trust is among the most needed qualities in the business world, as described by Peter Krass in his enlightened *Book of Leadership Wisdom*,[9] or Peter Cohan in his excellent writings on *Value Leadership*,[10] or Francis Fukuyama in his extraordinary book of *Trust*.[11] As essential as trust is for business leaders, it is equally essential for surgeons who are learning their craft or practicing their profession. Trust is indispensable for surgeons' patients and colleagues in order to build a world of confidence and understanding. In the same manner, David Packard (1912-1996) of Hewlett-Packard fame had a strong belief in people and dedicatedly argued for the unique value of trust in the life and work of others.[9]

What lessons can be learned by upcoming surgeons in the arena of trust? Is there a way for this quality or virtue to be continuously presented and spread in the walls of academia and surgical practice? Yes. I think the use of clear examples of trust in society, history, and literature are very valuable tools. I think the direct participation of surgeon teachers in alerting and stimulating young specialists to be trustworthy in their endeavors creates the necessary atmosphere to keep trust alive at all times. I think that emphasizing trust as a way of life and style of professional commitment is extremely helpful. I think trust is a lifetime virtue from which students and teachers alike derive their energy and enthusiasm in carrying out their noble careers.

References

1. Webster II New College Dictionary. Boston: Houghton Mifflin Company, 1999.
2. Small M. Being Trustworthy. Minneapolis: PictureWindow Books, 2006.
3. Toledo-Pereyra LH. Compassion. J Invest Surg 2005; 18:157-160.
4. Toledo-Pereyra LH. Respect. J Invest Surg 2005; 18:281-284.
5. Toledo-Pereyra LH. Integrity. J Invest Surg 2006; 19:1-3.
6. Toledo-Pereyra LH. Zabdiel Boylston. J Invest Surg 2006; 19:5-10.
7. Toledo-Pereyra LH. Vignettes on Surgery, History and Humanities. Georgetown, Texas: Landes Bioscience, 2005.
8. Earl AS. Surgery in America: From the Colonial Era to the Twentieth Century. New York: Praeger Publishers, 1983.
9. Krass P. Book of Leadership Wisdom. New York: John Wiley and Sons Inc., 1998.
10. Cohan PS. Value Leadership. San Francisco: Jossey-Bass, 2003.
11. Fukuyama F. Trust: The Social Virtues and Creation of Prosperity. New York: The Free Press, 1995.

CHAPTER 12

Gratitude

Luis H. Toledo-Pereyra

> *Gratitude is the mother of all virtues; without it, life loses all its meaning.*

A group of senior surgical residents, who in a few months would be finishing their general surgery residency program, were commiserating about the roughness of some of their surgical mentors. In the residents' own thinking, gratitude was distant since they basically equated gratitude to niceness and good treatment, which they felt they were not always receiving. In their own thinking, they were only considering the surgical faculty who were close to them. In their own thinking, faculty members were grateful but failed to express it appropriately. This was not something surgeons were prepared to do. After all, the residency program, at first glance, did not appear to be dedicated to enhancing the presence and value of gratefulness to improve both life and the world of surgery. In this writing, my dedicated intent is to emphasize the clear importance of gratitude in the professional lives of surgeons as well as in that of all people.

To write about gratitude is not difficult for me since I firmly believe in the great importance of exalting this significant and unique virtue.[1-18] For years I have felt as Seneca did that "nothing is more honorable than a grateful heart".[15] I have also overwhelmingly accepted the thinking of Cicero in that "gratitude is not only the greatest of virtues, but the parent of all the others".[15] Elie Wiessel, well-known Holocaust survivor and Nobel Peace Prize winner, put it very well when he said, "When a person doesn't have gratitude, something is missing in his or her humanity. A person can almost be defined by his or her attitude toward gratitude".[15] Furthermore, Sarah Ban Breathnach summarized it perfectly when she said, "Every time we remember to say thank you, we experience nothing less than heaven on earth".[11]

Reminiscences on Surgery, History and Humanities,
edited by Luis H. Toledo-Pereyra. ©2007 Landes Bioscience.

Gratitude, then, is what we need at home, at work, and in the operating room. Just imagine the great wonders of society and the world if gratitude could be bottled and used in all our endeavors, at all times, and under any circumstances. Gratitude, then, is the motor that makes engines function effectively and last a long time. Gratitude, then, is what we desperately need to improve our lives and to maintain the richness of our spirit. Gratitude, then, is a way of life, a heart filled with sentiment, a new spirit walking through the solitary meadows of southern Georgia, the force that brings about common sense and tolerance. Gratitude, then, is our guide, our strength, our hope, and our vision. Gratitude is all and for all.

Now let's be practical and see how we can teach surgical residents and faculty alike about the greatness of gratitude. Let's start by using a common example, one that we can readily imagine: the surgical resident and faculty not knowing and, for that matter, not acknowledging or explicitly thanking the important support staff who transport patients to the operating room suite. The surgical resident and faculty taking for granted the OR staff, nurses, surgical assistants, circulating personnel, and the anesthesia group. The surgical resident and faculty not communicating a great deal or not always clearly expressing their gratitude to the recovery room and regular floor staff. I guess the examples could multiply, even though they share a common denominator—the diminished emphasis on gratefulness and our attitude towards it.

Instead, let's assume for a minute that gratitude is frequently considered and utilized by surgeons and all other professionals in our society. Our responsibility and continuous awareness should be oriented towards maintaining gratitude and its related virtues in the minds and activities of each one of us. The task could be simple or complicated depending on the participating group. Our main requirement is to be part of the group.

Where are we going with all of this? Are we saying to the world, and among the surgeons and their group, that gratitude is not fully incorporated into surgical practice and daily professional activities? I do not think so. We are simply inviting surgeons, in this case, to be more aware of the attitudes of gratitude. As M. J. Ryan elegantly put it, "Consciously cultivating thankfulness is a journey of the soul, one that begins when we look around us and see the positive effects that gratitude create".[1]

Allow me to expand the knowledge of gratitude somewhat further. As you know, gratitude "is a feeling of thankfulness and appreciation at a benefit received. Feelings of gratitude can strengthen friendship and communion when accepted with an assumption of good faith".[2] Gratitude has been extensively studied from the psychological point of view in many ways. Specific areas such as the value of the benefit of gratitude, the cost to the benefactor, the intention of the benefactor, the gratuitousness of the benefit, and the determinants of gratitude are just a few of the parameters reviewed.[3] Ortony and his group[8] believed that the intention is at the basis of a highly regarded act of gratitude. Without intention, there is no real gratitude or there is no recognition of the virtue.

Benevolence is another important factor in understanding and enhancing the value of gratitude. Berger[3] reviewed this topic from a different perspective by adding the benefactor's benevolence in the act of gratitude. And he expressed that "the benevolent concern for the recipient will add to the value of a favor".[3] This is a very understandable and reasonable position when reviewing the value of this important action.

In summarizing the previous thoughts, we can consider that intention, benevolence, and spontaneity are key factors in appraising gratitude. The more intentionality we encounter, the more the worthiness of gratitude. The greater the benefactor's good intentions, the greater the expression of gratitude. The greater the spontaneity, the greater the feeling of gratitude. In short, gratitude needs to be felt as deliberately intended for the recipient and, of course, freely given.[1-10]

Another significant characteristic of gratitude is the "perceived cost of the benefit to the benefactor".[3] The greater the perceived cost, the more believable gratitude becomes. In addition, if the potential value of the benefit is clearly defined, the appearance of gratitude is readily identified and positively accepted. There is a clear relationship between the cost of the benefit and the recognition and highly regarded praiseworthy virtue of gratitude.

In returning to our surgical quest for securing the use of gratitude for all involved in the surgical arena, we must refocus our attention towards enhancing the presence and value of this great virtue. Planting the seed of gratitude in the surgical world is our highest desire. The specific plan for reaching many surgeons in practice, or future surgeons in training, calls for the development of a program that will highlight the eternal implications of gratitude. The plan is simple and complex at the same time. It is simple because the approach consists of practicing gratitude, "even if you don't feel it".[1] The more you do it, the more you will subsequently remember to incorporate its presence into your daily activities. It is complex because each individual responds in a different manner and creates his/her own attitude of gratitude. It appears at this point that a combination of the two approaches would be strongly recommended.

My humble advice to surgical fellows and colleagues alike would be to practice gratitude whenever you can throughout your professional and personal life. Use all moments of your career to enhance this enormous gift, the gift of gratitude. Take the goodness of this incredible joy into the lives of others. Give full humanness to your existence. Advance the lives of others by committing to appreciate their causes and function in their daily living.

Mark Nepo, a kind person, distinguished poet and philosopher, who recently spoke to our group, said it very well when he so clearly indicated that "as we are broken open by our experience, we begin to be grateful for what is, and if we live long enough and deep enough and authentically enough, gratitude becomes a way of life".[1] And this is the stage we want to reach. We want gratitude to be significant enough that it becomes a way of life.

Gratitude can be expressed at different levels, all of them of great importance. M.J. Ryan, in her outstanding book on *Attitudes of Gratitude*,[1] profoundly analyzes the levels of gratitude and defines them. She indicates that contentment, meaning, and joy are the three most significant levels existent in a good experience of gratitude. The degree to which they are represented constitutes the basis for improvement in professional and individual health.

Now let's return to the surgical residents in their informal meeting. They pondered the value of recognizing and honoring their teachers without recognizing gratitude itself, without considering the far reaching effects of gratitude. Tennis star Althea Gibson said it very well: "No matter what accomplishments you achieve, somebody helped you".[1] How truthful and how many times we take it for granted, how many times we don't look back and acknowledge our teachers, how many times we need to sit down and appreciate the great mentors who gave us the knowledge and opportunities we all enjoy today! "Appreciation makes immortal all that is best and most beautiful" indicated Percy Bysshe Shelley so wisely.[1]

Those of us who are convinced of the enormous personal and moral advantages of gratitude are responsible for teaching the benefits of this virtue to our colleagues and fellow residents alike. Gratitude builds confidence, promotes positive feelings, advances an encouraging atmosphere, and develops happiness in our work and private lives. As we spread the uplifting effect of this virtue, we should aim for an environment that promotes the "attitudes of gratitude," where surgeon staff, resident, and assistants are intimately involved in practicing the principles and essence of gratitude. I envision a world where surgical staff will always appreciate the dedicated commitment of fellow residents, where fellow residents will always respect and be grateful for the good deeds of the teaching faculty, and where both, resident and staff, will offer constant appreciation to everyone in the team. This is the world I see and admire. This is the world I would like to be part of. This is the best world for all.

References

1. Ryan MJ. Attitudes of Gratitude. York Beach: Conari Press, 1999.
2. Gratitude. Wikipedia, the free encyclopedia. (Accessed March 16, 2006. Available at http://en.wikipedia.org/wiki/Gratitude).
3. Berger FR. Gratitude. Ethics 1975; 85:298-309.
4. Emmons RA, Shelton CM. Gratitude and the science of positive psychology. In: Snyder CR, Lopez SJ, eds. Handbook of Positive Psychology. New York: Oxford University Press, 2002.
5. Lane J, Anderson NH. Integration of intention and outcome in moral judgment. Memory and Cognition 1976; 4:1-5.
6. McCullough ME, Kilpatrick SD, Emmons RA et al. Is gratitude a moral affect? Psychological Bulletin 2001; 127:249-266.
7. McCullough ME, Tsang J, Emmons RA. Gratitude in intermediate affective terrain: Links of grateful moods to individual differences and daily emotional experience. J Pers Soc Psychol 2004; 86:295-309.
8. Ortony A, Clore GL, Collins A. The Cognitive Structure of Emotions. Cambridge: Cambridge University Press, 1988.
9. Tesser A, Gatewood R, Driver M. Some determinants of gratitude. J Pers Soc Psychol 1968; 9:233-236.
10. Emmons RA, McCullough ME. Highlights from the Research Project on Gratitude and Thankfulness: Dimensions and perspectives of gratitude. (Accessed March 16, 2006. Available at http://psychology.ucdavis.edu/labs/emmons/).
11. Quotes to Live By: Gratitude. Colleen's Corner. (Accessed January 16, 2006. Available at http://colleenscorner.com/Gratitude.html).
12. Simpkinson AA. A gratitude revolution: Eight ways to make gratefulness. Beliefnet.com. (Accessed January 16, 2006. Available at http://www.beliefnet.com/story/93/story 9341 1.html).
13. The cultivation of gratitude and appreciation. To Do Institute's Resource Library. (Accessed March 16, 2006. Available at http://www.todoinstitute.org/gratitude.html).
14. Beliefnet Picks. Beliefnet.com. (Accessed January 16, 2006. Available at http://www.beliefnet.com/index/index 31990.html).

15. On gratitude. Beliefnet.com. (Accessed January 16, 2006. Available at http://www.beliefnet.com/story/53/story 5396 1.html).
16. Steindl-Rast D. Life Is a Gift. (Beliefnet.com. Accessed January 16, 2006. Available at http://www.beliefnet.com/story/51/story 5115 1.html).
17. Easterbrook G. Rx for life: Gratitude. (Beliefnet.com. Accessed January 16, 2006. Available at http://www.beliefnet.com/story/51/story 5111 1.html).
18. Spirituality and health: Gratitude. Spirituality and Health: The Soul/Body Connection. (Accessed January 16, 2006. Available at http://www.spiritualityhealth.com/newsh/items/blank/item 188.html).

12

CHAPTER 13

Innovation

Luis H. Toledo-Pereyra

> *The ability of someone to change old ideas, to modify*
> *well-established principles, and to challenge the ordinary.*

Peter F. Drucker (b. 1909), the talented Austrian-born writer, judicious observer, and teacher of management techniques, extensively reviewed the principles of innovation in *Innovation and Entrepreneurship*, published in 1985.[1] In this book, he emphasizes and analyzes knowledge-based innovation, and expands on the key ideas of advancement of knowledge and application of technology. In as much as he believes in the "unsuccessful attempts to identify the personal traits, behavior, or habits that make for a successful innovator",[1] he proposes simplicity and focused-oriented activities as essential principles associated with effective innovation.[1,2] We follow these ideas and search for better means to teach innovation as a way to maximize our ability to transform the ordinary in the surgical sciences into the extraordinary.

One of our advanced and thoughtful senior surgical residents had finished his daily rounds and was chatting with Dr. Mackey, his surgical career mentor. The young trainee had recently attended a national surgical congress where he incredulously witnessed the emphasis placed on the necessity of being a practical and common ground surgeon rather than an innovator. Even though it was clear to him that practicality and innovation were not antagonistic in principle, he sought the perspective of his admired mentor, surgeon Mackey. After hearing of this experience, Mackey deliberately turned to the resident with surprised eyes and exclaimed, "Innovation is everything! Let me explain what I mean. Not infrequently, you will hear that surgeons need to concentrate on what we do well and eliminate thoughts of grandiose, innovative ideas. What an incongruent, irrational, irritating thought! On the contrary, I believe surgeons need to think and practice innovation whenever possible. Innovation should

Reminiscences on Surgery, History and Humanities,
edited by Luis H. Toledo-Pereyra. ©2007 Landes Bioscience.

be part of our daily activities; innovation is everything!" The smart resident whole-heartedly agreed, finding his mentor's confirmation both reassuring and uplifting!

Now, the difficult task—how to teach innovation to surgical residents and faculty, how to exalt its virtues and how to proceed in the effective practice of innovation? The great enthusiasm of surgeon Mackey, admittedly a fictional character, should be contagious and worth imitating.[3] In the past, creative and innovative surgeons have been a frequent source of our writings. Indeed, we can refer to many inspiring examples, in particular to the *Nobel Laureate Surgeons*.[4,5,6] This dedicated group of individuals demonstrates a whole range of innovative abilities and exemplify a life of innovation.

For a brief moment, consider the unique qualities of these Nobel surgeons. Certainly, innovation is one of them together with commitment, determination, and focus.[5] These qualities, each important on its own, require the others to reach the potential climax of success so evidenced by these accomplished surgeons. Charles Huggins, one of the nine Nobel surgeons, would remind his students daily of the importance of discovery by asking, "What did you discover today?"[7]

From a practical point of view, what is the most effective way to teach innovation so everyone participates in the experience? How do we excel through innovation? How do we improve our professional practice by using innovative principles? There are no simple answers, but being part of the experiment justifies the effort. First, an annual course of theoretical lectures dealing with innovation should be instituted. Second, a series of books and papers about innovation, obtained from the business literature, should be included in the curriculum. Third, during rounds, in conferences, and in the operating room, innovation should be a topic of frequent discussion. Finally, residents should be required to write a paper or present a lecture dealing with innovation. If these suggestions were incorporated into residency programs, our residents would be better prepared to serve as the innovative surgeons of tomorrow. It is feasible that our residents would develop habits of change and improvement. It is feasible that our residents would be agents for enhanced operative techniques and surgical management.

Before concluding, we could receive some lessons from the greatest innovator of all time, Leonardo da Vinci (1452-1519). Can da Vinci teach us some practical and applicable lessons? Was he seeking innovation continually? Did he cultivate some principles that allowed him to innovate effectively? The most plausible answer is yes for each of these questions. Yes, da Vinci qualifies as a competent teacher of innovation, in as much as he pursued it throughout his lifetime. Yes, he cultivated solid principles to reach heights of excellence in innovation and pure knowledge.

Da Vinci was his own man, unafraid to challenge authority. He questioned old and new principles that did not conform to his expectations. His continual quest was to reach the truth. He was a consummate experimenter with wide-ranging curiosities. He acted based on knowledge and experience.[8-10] According to one of the great students of da Vinci, Michael Gelb, the master considered himself a *discepolo della esperienza* (disciple of experience).[10] Though da Vinci was not a scholar or academician, he occupied himself in gathering knowledge and actively searching for answers to critical questions. His guiding principles were independence, curiosity, originality, quest for knowledge, and determination. Da Vinci's life and virtues demonstrate how we can improve our innovative skills in daily professional practice.

Thomas Alva Edison (1847-1931) embodies the classical example of American innovation.[11-14] He and another distinguished American, Benjamin Franklin (1706-1790), were the most notable heroes of innovation on the American continent. Both of them produced many works of great significance. For this writing, I concentrate on the good deeds of "The Wizard of Menlo Park," as Edison was known in New Jersey where he lived.[11-14] Starting with the telegraph, the phonograph, the dictaphone, and the electrical lamp (incandescent light bulb), he completed 1093 successful patents in his lifetime, an unmatched record. Edison was an innovator's innovator all his life. He lived and breathed innovation. He was a practical man who saw needs and took advantage of the opportunities to fill them. His virtues were those of perseverance, full commitment, dedication, focused planning, and the desire to succeed at all cost. These significantly special characteristics distinguished Edison as "the innovator of the millennium".[12]

As a way to continue the extraordinary legacy of Thomas Edison, the Edison Preservation Foundation organizes the annual Edison Innovation Conference, where the Edison Innovation Awards are given to those especially gifted individuals who have pursued and demonstrated "Edison's legacy of entrepreneurship and innovation".[11-14] Given Edison's unique life and superb accomplishments, it is not hard to admire the man who is considered to be the "inventor of the 20th century" and the "father of modern invention".[12] Without doubt "his genius paved the way for the development of the computer, cell phone, and compact disk".[12,13] His brilliant example capitalizes on the best virtues a person can possess as a dedicated innovator.[1-16]

Like surgeon Mackey and the interested resident, we can ponder the potential benefits of this especially important and infrequently cultivated virtue of innovation. We, like they, attempt to define and characterize innovation as it differs from creation and discovery. One way to separate the three qualities of innovation, creation, and discovery is to consider innovation as an agent of change, creation as an agent of developing something new, and discovery as a mechanism to encounter something that was not known to exist before. In this form, you can be an innovator, creator, and discoverer simultaneously, or you can be an innovator and not necessarily a creator or discoverer. Each one, I believe, has its special characteristics and well-defined boundaries. Today, we are talking about the possibilities of incorporating only innovation into professional activities. Surgeon Mackey would claim there is much more to discuss. Still, he and his younger colleague could depart having explored new and enhanced possibilities for integrating innovation into surgical practice.

References

1. Drucker PF. Innovation and Entrepreneurship. New York: Harper Collins, 1985.
2. The Business World According to Peter F. Drucker. (Accessed May 30, 2006. Available at http://www.peterdrucker.com/).
3. Toledo-Pereyra LH. Integrity. J Invest Surg 2006; 19:1-3.
4. Toledo-Pereyra LH. Martinez-Mier G. Maestros Nobel de la Cirugia. Mexico: JGH Editors, 2000.
5. Toledo-Pereyra LH. Nobel Laureate Surgeons. J Invest Surg (In press).
6. Toledo-Pereyra LH. Vignettes on Surgery, History and Humanities. Austin: Landes Bioscience, 2005.
7. Toledo-Pereyra LH. Discovery in Surgical Investigation: The Essence of Charles Brenton Huggins. J Invest Surg 2001; 14:251-252.
8. Toledo-Pereyra LH. Leonardo da Vinci: The Hidden Father of Modern Anatomy. J Invest Surg 2002; 15:247-249.
9. White M. Leonardo: The First Scientist. New York: St. Martin's Griffin, 2000.
10. Gelb MJ. How to Think Like Leonardo da Vinci. New York: Delta Trade Paperback Edition, 2004.
11. Thomas Edison. (Accessed May 30, 2006. Available at http://en.wikipedia.org/wiki/Thomas Edison).

12. Edison Innovation Awards. (Accessed May 30, 2006. Available at http:/
    /www.edisoninnovationawards.com/).
13. McCormick B. At Work with Thomas Edison. Canada: Entrepreneur-
    ship Press, 2001.
14. Toledo-Pereyra LH. Lessons from Thomas Alva Edison—The Great-
    est American Inventor—To Surgical Investigators. J Invest Surg 2003;
    16:185-188.
15. Christensen CM. The Rules of Innovation. Tech Review 2002;
    105:32-39.
16. Peterson AJ. Leading the Way Across Generations. Med Bull Fall 2004;
    15-19.

13

CHAPTER 14

Loyalty

Luis H. Toledo-Pereyra

> *If gratitude is the mother of all virtues, loyalty is the most distinguished member of the family since loyalty is the most required of all of them.*

Marcus Aurelius Cavallini was a poised and extroverted junior surgical resident who knew at his young age how to treat colleagues and patients with elegance, superb caring, and visible compassion. He had arrived from Naples in southern Italy two years ago, and had adapted fairly well to American cultural standards. One thing was universal, Marcus Aurelius felt, the virtues of men, and of these, the practice of loyalty in particular.

Cavallini could not understand the fuss about loyalty since this virtue had been practiced for thousands of years, since the dawn of civilization. Sumerians, Egyptians, Greeks, and Romans had understood and embraced it since early times. "Those who have loyalty will conquer the world and bring respect and protection to those who deserve it," he thought, and then added, "Loyalty means understanding, loyalty represents caring for someone, loyalty is all about respect."

Many examples in the history of surgery represent the value and significance of loyalty in the private and professional life of the practicing surgeon. The loyalty of master neurosurgeon Harvey Cushing (1869-1939) to professor William Halsted (1852-1922) of Johns Hopkins stands without further consideration.[1] The respect, appreciation, and loyalty Cushing possessed and evidently felt for Halsted was legendary. Cushing was grateful to his teacher and above all loyal to all recommendations and beliefs of his mentor. This relationship continued throughout both of their lives, until the death of the great surgeon William Halsted, and thereafter until the erudite neurosurgeon Cushing passed away.[1]

Reminiscences on Surgery, History and Humanities,
edited by Luis H. Toledo-Pereyra. ©2007 Landes Bioscience.

Similar loyalty evidently existed between the gifted disciple Walt Lillehei (1918-1999) and his accomplished surgeon-teacher Owen Wangensteen (1899-1981) of Minnesota. They had a unique relationship, one that reached the highest levels of intimacy and respect, one that was censored by loyalty and admiration and crowned by a level of respect usually reserved for fathers and sons. Wangensteen, the chief, as he was respectfully called, was always there for his innovative student. At no moment when the professor was needed did he dismiss keen participation to help his student. Lillehei reciprocated the same feelings towards his distinguished mentor and dedicated teacher. This strong sense of loyalty actively continued after professor Wangensteen's death and persisted until Walt Lillehei's death. I remember the noted cardiovascular surgeon on my visit to his private residence at 93 Otis Lane in St. Paul. He reminded me of the enormous depth of gratitude he had for his dear and esteemed mentor, and particularly emphasized the intense loyalty that characterized their relationship. This loyalty remained vibrant throughout his life, firmly took hold, and appeared to reach higher levels in old age. As I see it, the unique Wangensteen-Lillehei combination of respect, consideration, and loyalty transcends as a good example for future generations of surgeons to imitate and pursue enthusiastically.

In Chicago, surgeons Charles Huggins (1901-1997) and Dallas Phemister (1882-1951) had a strong relationship built on trust and loyalty.[2] The earnest new staff and disciple (Huggins) had cultivated his association with the recognized teacher (Phemister) through intense work ethics and respect for the assimilation of the ideas and suggestions of the chief surgeon. When Huggins had to decide about his future surgical specialty, to which he would dedicate his entire life, he followed the wise words of Dr. Phemister, who recommended attending the most important European urological clinics available at the time. There Huggins experienced the great sense of the need for this discipline and become a urological surgeon with excellent knowledge.[2] The young and future urologist followed with dedication the advice of his Chicago professor, accepting guidance from someone whom he admired and utterly respected. Certainly, loyalty was at the center of their very special relationship. Teacher and staff surgeon protected their relationship through the years, allowing room for understanding and a caring attitude.

More recently, I have seen for 36 years the development of a special relationship of loyalty and trust, consistently growing between a disciple and master surgeon at Minnesota. David Sutherland

(b. 1940), trusted friend and distinguished surgeon, and John Najarian (b. 1927), accomplished professor and respected surgeon-in-charge for many years, have cultivated solid bonds of respect, appreciation, and loyalty. Fame and personal preferences have not distracted them from maintaining a close union in accepting and following each other's ideas and accomplishments. Najarian and Sutherland, through mutual respect and loyalty, have handsomely surpassed any difficulties that could have appeared during their 36 years of productive and uplifting professional and personal relations. Their association remains as a clear and vivid example of how teacher and student can enrich their own lives and those around them by observing essential, natural principles of understanding, loyalty, and gratitude.[3-14] Sutherland and Najarian first encountered each other in Minneapolis and have maintained high levels of friendship and trust throughout their lives.

The vivid examples discussed before clearly represent the value and importance of trust and loyalty.[1-14] Both of these virtues are the best companions to the life and meaning of human beings. The surgical experience is not an exemption to the incorporation of these principles into personal life and professional practice. The reality is that surgeons are like anyone else; they can have a high level of loyalty or can be devoid of this great virtue. We need, therefore, to cultivate this human quality. How can we organize a system that creates a sustainable plan for recognizing and stimulating loyalty among colleagues and friends? Not a simple task but one that is doable, I believe.

The quest for loyalty commences at an early stage, when residents and fellows are beginning their training and when exposure to new faculty and ways of treatment starts shaping their new profession. This is the time when we need to expand the benefits of loyalty and incorporate it prominently into everyone's thinking. This it the time when good examples are critical for young physicians to learn the chief virtues of humanity and how to use them more effectively. This is the time for clear understanding of loyalty and how to learn its basic principles.

Loyalty is important at all moments of training and professional life.[1-14] Loyalty is essential during the planning of treatment. Loyalty is needed in the management of patients and the appropriate use of the best surgical techniques. Loyalty represents the ideal way to establish trust and create an improved relationship with colleagues and faculty mentors.

A former Japanese surgical fellow, working with me at the Borgess Surgical Research Laboratories in the early 1990s, exalted *The Eight Principles of Bushido*[15,16] as the main road to successful living and to remaining loyal to colleagues and masters. These were the same principles maximally identified by the Samurai, an extraordinary group of ancient warriors, the same principles proclaimed as truth by the same Samurai class of fighters, and the same principles utilized by those who wanted to live in virtue and honor.[15,16] Since loyalty plays such a crucial role in the life of people practicing Bushido, it would be helpful to include here all of the principles associated with this way of living.[15,16]

The Eight Principles of Bushido

1. Jin—to develop a sympathetic understanding of people
2. Gi—to preserve correct ethics
3. Chu—to show loyalty to one's master
4. Ko—to respect and to care for one's parents
5. Rei—to show respect for others
6. Chi—to enhance wisdom by broadening one's knowledge
7. Shin—to be truthful at all times
8. Tei—to care for the aged and those of a humble station

It has been said that old-fashioned loyalty is not easy to reproduce in our current society. From the 15th century when the word came into use, the meaning of *loyalty* was associated with "sense of fidelity" to cause, love, family, friends, work, and other professional activities.[5] Being loyal meant being supportive of someone's cause. Being loyal meant being faithful to another's way of pursuing life and work endeavors. Being loyal meant total commitment to someone's principles of justice and understanding.

I can recommend speaking and practicing loyalty whenever possible, and advise pursuing the principles followed by loyal individuals. Even Marcus Aurelius Cavallini, a good fictional character, serves as a keen example of trust and loyalty. It is not hard to understand that the application of this virtue is particularly important to those who provide education and training. Indeed, surgeons have a unique opportunity to excel in loyalty when teaching in rounds and in the operating room. Many surgeons are the best source of caring and support. Many surgeons practice loyalty during teaching, daily practice, and in personal endeavors. They are, in fact, the best role models to be followed by younger faculty, residents, and fellows alike.

References

1. Toledo-Pereyra LH, ed. Vignettes on Surgery, History and Humanities. Austin: Landes Bioscience, 2005.
2. Toledo-Pereyra LH. Nobel Laureate Surgeons. J Invest Surg 2006; 19:211-218.
3. The quest for loyalty. (Accessed July 17, 2006. Available at: http://pubs.acs.org/hotartcl/chemtech/96/dec/quest.html).
4. Stanford discovers why business loyalty programs work. Silicon Valley/San Jose Business Journal Online. (Accessed July 17, 2006. Available at: http://sanjose.bizjournals.com/sanjose/stories/2004/08/16/daily2.html).
5. Loyalty. Wikipedia. (Accessed May 2, 2006. Available at: http://en.wikipedia.org/wiki/Loyalty).
6. Loyalty business model: Wikipedia. (Accessed July 17, 2006. Available at: http://en.wikiedia.org/wiki/Loyalty_business_model).
7. Johnson LK. Rethinking company loyalty: Harvard Business School Working Knowledge. (Accessed July 17, 2006. Available at: http://hbswk.hbs.edu/item/5000.html).
8. Loyalty rules! Harvard Business Review. (Accessed July 17, 2006. Available at: http://www.loyaltyrules.com/loyaltyrules/library_harvard.html).
9. The quest for loyalty. Chemtech 1996; 26(12):14-18, (Accessed July 17, 2006. Available at: http://pubs.acs.org/hortartcl/chemtech/96/dec/quest.html).
10. Toledo-Pereyra LH. Trust. J Invest Surg 2006; 19:69-71.
11. Toledo-Pereyra LH. Integrity. J Invest Surg 2006; 19:1-3.
12. Toledo-Pereyra LH. Gratitude. J Invest Surg 2006; 19:137-140.
13. Toledo-Pereyra LH. Respect. J Invest Surg 2006; 18:281-284.
14. Toledo-Pereyra LH. Compassion. J Invest Surg 2005; 18:157-160.
15. The Samurai and their use of Bushido. (Accessed August 3, 2006. Available at: http://victorian.fortunecity.com/duchamp/410/bsamurai.html).
16. Alexanian MG. The seven principles of Bushido—The way of the warrior. (Accessed August 8, 2006. Available at: http://ejmas.com/tin/tinart_alexanian_0402.htm).

14

Section III.
Surgeons, Pioneers, Educators

CHAPTER 15

Zabdiel Boylston

Luis H. Toledo-Pereyra

The Boston of Old State House, the Boston of Old South Meeting Place, the Boston of the first public school, the Boston of Old North Church, and the Boston of Beacon Hill were all frequently attended and specially considered in the life and activities of notable citizen surgeon Zabdiel Boylston. Brookline, then known as Muddy River, was the origin of this unique American surgeon pioneer.

Zabdiel Boylston grew up in an atmosphere of abundance and support. His father was the prominent doctor Thomas Boylston (1644-1695), a well-respected physician and chirurgeon from Muddy River, who supposedly obtained his medical degree from Oxford. His mother was Mary Gardner, a distinguished lady, the daughter of Thomas Gardner and Lucy Smith, also from Muddy River. The Boylstons and Gardners were well-known families who had come from England just a few years earlier.[1-9]

Boston was founded in 1630 by a group of courageous Puritan colonists. The colony of Massachusetts was progressing at an accelerated pace and, by the late 1600s and early 1700s, included many of the special advances seen in Europe at this time.[10] Higher education in the American colonies had not reached the level of some European nations; for instance, professional schools had not yet been integrated. In fact, the first American medical school in the English colonies was not founded until 1765 at the University of Pennsylvania. Columbia and Harvard came immediately thereafter.

Zabdiel Boylston learned medicine from his father, Thomas Boylston, and another eminent Boston physician, John Cutter.* After several years of apprenticeship, the younger Boylston started the practice of medicine and surgery. He began at a slow pace in the city of Boston and carefully applied the principles he'd so diligently absorbed from his earnest father and the advanced Dr. Cutter. Boylston was fifteen when his father passed away of an unrecorded cause. There-

*Genevive Miller recognizes this name as Cutler.[9]

Reminiscences on Surgery, History and Humanities,
edited by Luis H. Toledo-Pereyra. ©2007 Landes Bioscience.

after, he progressed well under Cutter's tutelage. By the time he decided to practice medicine by himself, he was probably in his twenties. He married Jerusha Minot on January 18, 1705. By then, most likely, he was in practice. They had eight children together.[5,9]

Zabdiel Boylston had a successful family life and attained all his professional goals by the 1720s when he was in his 40s. He retired in his late 50s and died when he was 86 years old. He was a dedicated physician with enormous energy and determination. He attended the most difficult cases at the most inhospitable locations and whenever necessary, night or day. He did not travel by carriage, but rode a horse named Prince. Patients and the public alike recognized Dr. Boylston riding his stallion through town to attend his daily patient rounds. He was a doctor on horseback who never failed those patients in need.[11]

The first recorded surgery in the English American colonies was the operation performed by Zabdiel Boylston which consisted of the operative removal of a bladder stone on the son of Henry Hill in the city of Boston on June 24, 1710.[12-14] Other similar surgeries of operative bladder stone removals had been performed by surgeon Boylston but never registered, as occurred with this case. The *Boston News-Letter* on July 17-24, 1710, carried the first account of a published surgical case in colonial English North America:[12-14]

> *"For the benefit of any that has or may have Occasion, Henry Hill Distiller in the Town of Boston New-England, having had a child grievously afflicted with the Stone, apply'd himself to Mr. Zabdiel Boylstoun of the said Boston, Practitioner of Physick and Chirurgy; who on the 24th of June last, in presence of sundry Gentlemen, Physicians and Chyrurgeons, Cut the said Child & took out of his Bladder a stone of considerable bigness and with the blessing of God in less than a months time has perfectly Cured him, and holds his Water: This is his third Operation performed in the Stone on Males and Females, and all with good success: He likewise pretends to all other Operations in Surgery. Which operation the said Hill could not omit to make Publick."*

The second registered surgical contribution in the English American colonies belonged to Zabdiel Boylston as well. This was the operative excision of a breast tumor, most likely of cancerous origin. He swiftly removed the whole breast of Sarah Winslow in a matter of a few minutes. This operation occurred on July 30, 1718, in the presence of ministers and other attendants. The same newspaper in which the first case had appeared, the *Boston News-Letter,* on October 14, 1720, carried the following report of the second account of a published surgical case in English colonial North America:[13-15]

> *"For the public good of any that have or may have cancers: These may certify that my wife had been laboring under the dreadful distemper of a cancer in her left breast for several years; although the cure was attempted by several doctors from time to time, it was without success. When life was almost despaired of by reason of its repeated bleedings, growth, and stench, and she seemed to be in danger of immediate death: We sent for Dr. Zabdiel Boylston of Boston, who on July 30, 1718 (in the presence of several ministers and others assembled on that occasion), cut her whole breast off and dressed it in the space of five minutes by the watch of one then present; and by the Blessing of GOD on his endeavors, she has long since obtained a perfect cure. I deferred the publication of this so long lest it should have broken out again."*
>
> —Edward Winslow
> Rochester, Oct. 14, 1720

Zabdiel Boylston was a busy clinical physician and surgeon. Clearly, he had the utmost respect of the Boston community. His word was definitely heldstrongly against any other medical doctor in the region. Approximately less than fifteen doctors were practicing in Boston at the time. All of them had obtained their medical designation through apprenticeship and not from formal medical school studies, as there were no medical schools available in the American English colonies. The only exception was doctor William Douglass, a Scottish physician who had received medical education at various European universities. It is unknown whether another Boston physician, Dr. Dalhonde from France, had received any European education. Both of these physicians and many of their contemporaries were later on to vehemently oppose Zabdiel Boylston for his practice of inoculation.[15-23]

As an accomplished surgeon of the time, Zabdiel Boylston had excelled in being extremely careful, offering appropriate tissue protection, and being aware of the anatomic considerations of the operative site, particularly when lack of anesthesia and asepsis were the rule. Surgeons were required to demonstrate great dexterity and efficient turn-around. Time was essential to minimize pain and maximize patient tolerance. There were no standards for breast surgery in the early 18th century. Anesthesia and asepsis were more than 150 years away. Under these circumstances, the surgical genius of Zabdiel Boylston became even more and uniquely prominent.

There are other issues worth considering in the case of Mrs. Winslow pertaining to the wide excision of the breast tumor.[15] First, the patient was too sick to travel to Boston from her home at Cape Cod. Doctor Boylston and his assistant, Jack, a black slave who worked for him for many years, went to her home to perform the surgery. Second, most of the time was spent in cleaning the table, the floor, the hearth, the instruments, and finally, and as importantly, Dr. Boylston had instructed the maid to clean the patient's body and operative site. Third, he used alcohol in its various forms, such as strong rum, to scrub the patient's skin. And fourth, he carefully studied the operative site and gathered the help to hold Sarah down. The surgery was finished in a matter of minutes!

Doctor Zabdiel Boylston had an accurate idea of what the Boston public would think of such surgery. He knew his detractors would consider him "a butcherous quack" and his supporters would call him "daring, courageous, kind." There was no easy way out. He knew he needed to help Sarah Winslow and that is what he and Jack did on that July morning of 1718. In The SpeckledMonster, Jennifer Lee Carrell, in a very elegant manner, describes some of the characteristics of this extraordinary case:[15]

> *"Dr. Boylston had not previously advertised this spectacular case. Partly because he was not sure Sarah Winslow would live: cancers as far gone as hers often come back. Partly because no one in New England had ever dreamed of such an operation, much less performed one. He knew what would happen. Through most people's minds, over and over, would slice the vision of a circle of men holding a woman down on her own kitchen table, while he, Dr. Boylston, stood in the middle, one hand forking the diseased breast, the other scything through it with a ruthless blade. The flesh would fall to the floor as he reached for the red-hot cauterizing iron, sealing the wound by searing it, the sizzle and steam pierced by screaming of a kind rarely heard outside childbirth or war.*
>
> *That was the swift horror of surgery. What the doctor could not make most people understand was that it was also the most*

compact condensed part of an operation, lasting mere minutes.
The vast bulk of time was spent cleaning, not cutting. "

It is very much likely that Zabdiel Boylston continued to perform mastectomies after the case of SarahWinslow, principally because his son, Thomas Boylston (1715-1750), who had become a doctor under his father's apprenticeship, was trusted enough to perform a mastectomy of his own in 1737 in Newport, Rhode Island.[15] Thomas Boylston could have only learned the procedure from his father. This instruction must have occurred much after the case of Sarah Winslow, since in 1718 the younger Boylston would have been three years of age. It is unfortunate that Thomas Boylston died at such a young age. He might have accomplished a great deal more. When the young Boylston visited London and in particular St. Thomas Hospital, his father identified him to his dear friend, Sir Hans Sloane, as "the first fruit of inoculation in the AmericanWorld".[15] Indeed, what a substantial advancement!

Of significant consideration in documenting the surgical feats of Zabdiel Boylston is the historical scholarship of Francisco Guerra, who very diligently compiled in his *American Medical Bibliography*, published in 1962, the surgical operations performed in English early America.[14] An American surgeon historian, Scott Earle, encountered the work of Francisco Guerra and published it in 1965.[13] In this way, the whole circle of Zabdiel Boylston's contributions to surgery became recognized in the annals of surgical history.

Pioneering bladder and breast surgeries were followed by the third great medical and surgical contribution of Zabdiel Boylston-the introduction of smallpox inoculation on June 26, 1721.[5,9,15] Some consider this the greatest contribution to New World medicine. Zabdiel Boylston inoculated that summer morning his son Thomas, six years of age, together with two black slaves. All of them survived the inoculation and outlived the smallpox epidemic without difficulties. This simple surgical scarification of the arm with live smallpox virus constituted the beginning of a new era in the control of infectious diseases.

Many were the difficulties that Zabdiel Boylston encountered in the incredulous and jealous world of the Boston medical society of the 1720s, particularly in reference to smallpox inoculation. The good deeds all began with Cotton Mather (1663-1728), a well-informed and erudite minister.[24] He pointed out to Boston physicians a practice of inoculation for smallpox in Africa and the East (Turkey, China,

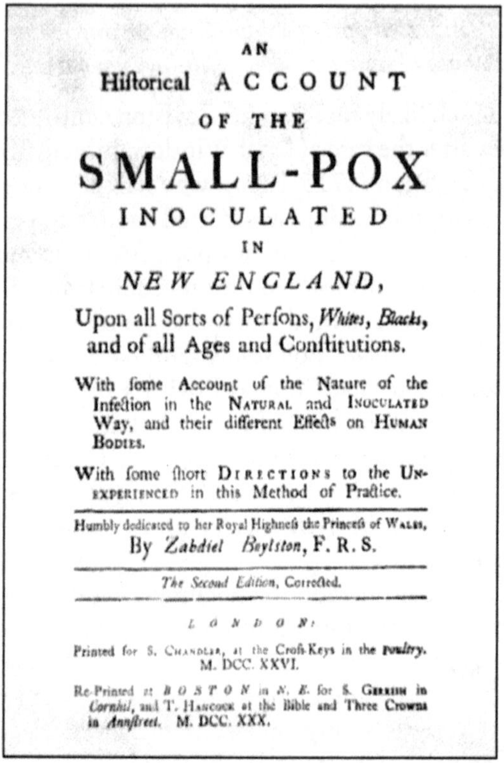

AN

Hiſtorical A C C O U N T

OF THE

SMALL-POX

INOCULATED

IN

NEW ENGLAND,

Upon all Sorts of Perſons, *Whites, Blacks,*
and of all Ages and Conſtitutions.

With ſome Account of the Nature of the
Infeſtion in the NATURAL and INOCULATED
Way, and their different Effeſts on HUMAN
BODIES.

With ſome ſhort DIRECTIONS to the UN-
EXPERIENCED in this Method of Practice.

Humbly dedicated to her Royal Highneſs the Princeſs of WALES,

By *Zabdiel Boylston,* F. R. S.

The Second Edition, Correſted.

L O N D O N:

Printed for S. CHANDLER, at the Croſs-Keys in the *Poultry.*
M. DCC. XXVI.

Re-Printed at *B O S T O N* in N. E. for S. GERRISH in
Cornhil, and T. HANCOCK at the Bible and Three Crowns
in *Annſtreet.* M. DCC. XXX.

Figure 1. Zabdiel Boylston's 1730 publication on smallpox inoculation.

etc.) recently introduced to him by Onesimus, his African slave, and also published in the *Transactions of the Royal Society* (1714).[15,24] "This communication was received with great contempt by all physicians in the Boston area except by Zabdiel Boylston who with intense energy and dedication proceeded to move ahead with smallpox inoculation. He immediately began inoculating in spite of severe and persistent opposition by doctorWilliam Douglass, 'a man of narrow mind and malignant passions' ".[15] Zabdiel Boylston persevered, and personally inoculated 247 individuals between 1721 and 1722, of which only six died, approximately 3%. The death toll for the general population infected with smallpox was 14%, and in some quarters, especially for children and the elderly, reached 50%.[15,16] Improving the smallpox survival rate by 11% qualifies as a remarkable achievement! (Fig. 1).

As a consequence of this life-saving procedure, Dr. Boylston received in London the highest accolades, including admission to the Royal Society. He became the first American to belong to this illustrious group of innovators and discoverers, mainly from the English-speaking world. At the request of Sir Hans Sloane, president of the Royal College of Physicians of England in 1721, Zabdiel Boylston enthusiastically and wishing to convey the message of inoculation to Europe, departed for London for an extended two-year visit (1724-1726).[5,7,9,25] His trip was expectedly well-received by staff, physicians, medical organizations, and extremely well-respected learned societies.[5,7,9] Zabdiel Boylston was the toast of the city, day after day.

Dr. James Jurin, secretary of the Royal Society, introduced Zabdiel Boylston to the medical and scientific community of London, and proved Boylston's best ambassador during the successful outreach trip. James Jurin, in many ways more than Sloane, took care of Zabdiel Boylston's needs while in the English capital.[5,7,9] In 1726, while in London, Boylston published his seminal work on the account of smallpox in New England, one that represented a classical work of his times and our times as well.[26,27] He dedicated this work to Caroline, Princess of Wales, though why is unclear. We know he most likely communicated with the Princess while in England, and we also know that she consistently remained interested in smallpox, as a mother of a young child (Princess Anne), who might contract the potentially deadly disease. Perhaps for Zabdiel Boylston these were good enough reasons!

In April 1721, two months before Zabdiel Boylston's inoculation in America, Charles Maitland had inoculated Lady Mary Wortley Montague's daughter in England. This was the first inoculation in Europe.[15,16] However, this significant development was never known to Zabdiel Boylston or Cotton Mather prior to the American inoculation. Mather had obtained the same information through other sources, while Lady Montague had secured similar information from the Royal Society (1716) through Dr. Pylarini from Smyrna, Turkey, where her husband was the English ambassador.[15,28] Another publication of Dr. Timonius, from Constantinople, had appeared two years before (1714) in the *Transactions of theRoyal Society*. So a trip that began in Constantinople and Smyrna had successfully circled to Boston. A trip of hope and encouragement. A trip of continuous support for those suffering the disease or those who potentially could be killed by smallpox.

Zabdiel Boylston "was remarkable for his skill, his humanity, and close attention to his patients".[5] He "possessed a strong and reflective mind, and acute discernment. His character through life was one of unimpeached integrity. He was charitable in his opinions of others, patient under the severest prosecution, and forgiving of his bitterest enemies".[11] In addition, "He was always faithful to his word".[11] In essence, Zabdiel Boylston represented the highest virtues of the surgical profession and of the human race as a whole!

References

1. Thomas Boylston. (Available at: http://www.wizard.net/_aldonna/tboylston2.htm. Accessed November 3, 2003).
2. Boylston, Zabdiel. The Columbia Encyclopedia. 6th ed. 2001, (Available at http://www.bartleby.com/65/bo/Boylston.html. Accessed November 3, 2003).
3. Boylston Zabdiel. (Available at: http://www.infoplease.com/ce6/people/A0808619.html. Accessed November 18, 2002).
4. Elliott CA. Biographical Dictionary of American Science. Westport, Connecticut: Greenwood Press, 1979.
5. Fulton JF. Boylston, Zabdiel. In: Johnson A, ed. Dictionary of American Biography, Volume 1. New York: Charles Scribner's Sons, 1964.
6. Fitz RH. Zabdiel Boylston, Inoculator, and the epidemic of smallpox in Boston in 1721. Bull Johns Hopkins Hospital 1911; 22:315-327.
7. Viets HR. Some Additions to a Biography of Zabdiel Boylston. In: Smit P, Laage RJ ter, eds. Essays in Biohistory. Utrecht: 1970.
8. Biographical Sketch of Dr. Boylston. Philadelphia: 1815.
9. In: Garraty JA, Carnes MC, eds. Boylston, Zabdiel. American National Biography. New York: Oxford University Press, 1999.
10. The World Book Encyclopedia, Volume 2. Chicago: World Book, Inc., 1985:403-412.
11. Thacher J. American Medical Biography, Volume 1. New York: Da Capo Press, 1967:185-192.
12. Rutkow IM. Zabdiel Boylston and the Earliest Published Account of an Elective Surgical Operation in Colonial America. Arch Surg 2002; 137:227.
13. Earle AS, ed. Surgery in America: From the Colonial Era to the Twentieth Century, Selected writings. 2nd ed. New York: Praeger, 1983.
14. Guerra F. American Medical Bibliography 1639-1783. New York: Lathrop C. Harper, 1962:463:450.
15. Carrell JL. The Speckled Monster: A historical tale of battling smallpox. New York: Dutton, 2003:99-141, 156-169, 170-189, 278-295, 395-424.
16. Hopkins DR. The Greatest Killer: Smallpox in history. Chicago: The University of Chicago Press, 2002:249-253.

17. Friedenberg ZB. The Doctor in Colonial America. Danburry, Connecticut: Rutledge Books, 1998.
18. Wallach G. The Mind of the Colonial Physician. Connecticut Medicine 1976; 40:815-827.
19. Sloan AW. English Medicine in the Seventeenth Century. Durham, North Carolina: Durham Academic Press, 1996.
20. Mager GM. Zabdiel Boylston: Medical pioneer of colonial Boston. Urbana-Champaign, Illinois: University of Illinois, 1975.
21. Smallpox in Colonial America. New York: Arno Press, 1977.
22. Goler RI. A household and its doctor: A case study of medical account books in Colonial America. In: Benes P, Benes JM, eds. Medicine and Healing. Boston: Boston University, 1992.
23. Duffy J. Epidemics in colonial America. Baton Rouge: Louisiana State University Press, 1953.
24. Silverman K. The Life and Times of Cotton Mather. New York: Columbia University Press, 1985:337-344.
25. Brooks ESJ, Sloane H. The Great Collector and his Circle. London: Batchworth Press, 1954.
26. Smit P, ter Laage RJCV, eds. Biographical Sketch of Dr. Boylston. Essays in biohistory and other contributions presented by friends and colleagues to Frans Verdoorn on the occasion of his 60th birthday. Utrecht, Netherlands: International Association for Plant Taxonomy, 1970.
27. Boylston Z. An Historical Account of the Small-Pox Incoculated in New England. London: S. Chandler, 1726.
28. Klebs AC. Historic Evolution of Variolation. Bull Johns Hopkins Hospital. 1913; 24:69-83.

CHAPTER 16

Ephraim McDowell

Luis H. Toledo-Pereyra

Humility and educated manners abounded in the character and personality of Ephraim McDowell (1771-1830). Of Virginian stock and Scottish/Irish ancestry, he spent the majority of his professional life in Danville, Kentucky.[1,2] In this town, he realized the incredible surgical feat of successfully removing a fifteen pound ovarian tumor from Jane Todd Crawford in 1809.

The conditions of the time did not permit for the use of general anesthesia, systematic or effective antiseptic practice, control of infection, or appropriate management of pain. Essentially, all the requisite elements of modern surgery were inexistent by a long shot. Surgeons were not performing abdominal operations, and only traumatic injuries would force them to delve into the suffering abdomen. Here is where McDowell advanced the practice of surgery to a level not seen before. No one else before him had ventured into the abdomen in a safe and practical manner. McDowell was a true pioneer in this regard, the first one to successfully and surgically navigate the human abdomen, and because of this, he should be considered the "father of abdominal surgery" (Fig. 2).

From his early beginnings, Ephraim dedicated himself to studying medicine as an apprentice under Dr. Alexander Humphreys of Staunton, Virginia.[1-3] Humphreys was an Edinburgh graduate who continued the Scottish tradition of teaching anatomy and surgery through dissection, with less emphasis on book academics. Ephraim learned through these principles, and after his apprenticeship, he followed in the steps of his mentor by attending Edinburgh medical school from 1792 through 1794. The teachings of Dr. John Bell were an important part of his education there. Even though he did not obtain an MD degree at the school due to a lack of personal funds, he was fully prepared to practice medicine upon his return to Danville, where he began his practice in 1795. When his practice reached stability, he married Sarah Shelby, daughter of the first governor of Kentucky. Together, they had six children.[1-3]

Reminiscences on Surgery, History and Humanities,
edited by Luis H. Toledo-Pereyra. ©2007 Landes Bioscience.

Figure 2. Ephraim McDowell.

Before taking Mrs. Crawford to surgery in Danville, McDowell had traveled on horseback for about sixty miles to evaluate this patient at her home. After full observation and examination, McDowell's diagnosis was a possibly enlarged ovarian cyst.[3] His diagnosis proved to be correct. He requested the family bring the patient on horseback to his office to proceed immediately with the case.[3]

To understand the extraordinary odyssey that the surgical case of Jane Todd Crawford represented, let McDowell tell us in his own words how the event evolved in 1809:[4]

> *"With the assistance of my nephew and colleague, James M'Dowell, MD, I commenced the operation, which was concluded as follows: Having placed her on a table of the ordinary height, on her back, and removed all her dressing which might in any way impede the operation, I made an incision about three inches from the musculus rectus abdominis, on the left side, continuing the same nine inches in length, parallel with the fibres of the above named muscle, extending*

into the cavity of the abdomen, the parietes of which were a good deal contused, which we ascribed to the resting of the tumor on the horn of the saddle during her journey. The tumor then appeared full in view, but was so large that we could not take it away entire. We put a strong ligature around the fallopian tube near to the uterus; we then cut open the tumor, which was the ovarium and fimbrious part of the fallopian tube very much enlarged. We took out fifteen pounds of a dirty, gelatinous looking substance. After which we cut through the fallopian tube, and extracted the sack, which weighed seven pounds and one half. As soon as the external opening was made, the intestines rushed out upon the table; and so completely was the abdomen filled by the tumor, that they could not be replaced during the operation, which was terminated in about twenty-five minutes. We then turned her upon her left side, so as to permit the blood to escape; after which we closed the external opening with the interrupted suture, leaving out, at the lower end of the incision, the ligature which surrounded the fallopian tube. Between every two stitches we put a strip of adhesive plaster, which, by keeping the parts in contact, hastened the healing of the incision...In five days I visited her, and much to my astonishment found her engaged in making up her bed. I gave her particular caution for the future; and in twenty-five days, she returned home as she came, in good health, which she continues to enjoy.[4]

The success of this case brought McDowell recognized fame. He did not stop there, but performed a total of 13 operations for ovarian cysts with full recovery of eight patients (63%).[3] Other surgeries were also part of his armamentarium such as herniorrhaphies and lithotomies. James Polk, the future United States president, was a patient of McDowell and underwent successful removal of a bladder stone.[3]

McDowell was not a dedicated writer. In fact, the publication of his first three cases did not appear until eight years after the first case.[3,4] He had sent his report to his former teacher, John Bell, and to the distinguished Philadelphia professor, Philip Physick. Since no response was obtained, nephew William took the paper to Thomas James, editor of the *Eclectic Repertory and Analytical Review*, who demonstrated great interest and published the case quickly thereafter.[4] Other cases followed. In spite of these reports, this important discovery was left untouched. The medical profession conceded no significance to these cases, and surgeons were not convinced of the positive effects of this technique.[3]

Years later, other Americans began using McDowell's technique of ovariotomy, now referred to as oophorectomy. In 1821, Nathan Smith was the second to perform this procedure.[6] Gross believed Smith knew McDowell's operation.[1,6] Alban Goldsmith, David Rogers, and J. Bellinger performed similar operations from 1823 onward. This operation then fell into disuse until the 1840s when John Atlee (1799-1885) and his brother, Washington Atlee (1808-1887), began their series of successful oophorectomies. John did 78 oophorectomies with 64 recoveries (82%). Washington performed 387 operations.[6] Charles Clay was carrying out similar successes in England.

By the mid-1950s, the terrain in North America was becoming more sympathetic towards this operation. Hostility and criticism were no longer apparent, and the name and deeds of McDowell were beginning to be accepted and praised. Monuments were erected in his memory, and surgeons all over recognized his unique and extraordinary contribution to abdominal surgery while still handicapped by the absence of asepsis and anesthesia. Without a doubt, McDowell well deserves the title of "father of abdominal surgery."

References

1. Gross SD. Lives of Eminent American Physicians and Surgeons of the Nineteenth Century. Philadelphia: Lindsay and Blakiston, 1861:207-230.
2. Horine EF. The Stagesetting for Ephraim McDowell, 1771-1830. Bull Hist Med 1950; 24:149-160.
3. Talbott JH. A Biographical History of Medicine. New York: Grune and Stratton, 1970:269-271.
4. McDowell E. Three Cases of Extirpation of Diseased Ovaria. Eclectic Repertory and Analyt Rev 1817; 7:242-244.
5. McDowell E. Observations on Diseased Ovaria. Eclectic Repertory and Analyt Rev 1819; 9:546-553.
6. Packard FR. History of Medicine in the United States. New York: Hafner Publishing Company, 1963:1130-1136.

CHAPTER 17

John Collins Warren

Alexander H. Toledo

John Collins Warren (1778-1856) embodied the essence of Boston surgery during the first half of the nineteenth century.[1-9] His father, John Warren (1753-1815), was a leading medical educator, surgeon, and patriot of eighteenth century Boston. Warren's father had the foresight to participate in founding Harvard Medical School in 1782, at the time the third such school in the English American colonies, after Pennsylvania and King's College (Columbia) in New York. In spite of John Warren's clear commitment to medicine, he did not recommend that his son follow in his footsteps. He considered a career in business to be more conducive to a successful and financially stable life.

After finishing at Harvard College in 1797 with a Bachelor of Arts, John Collins Warren pursued a business career by dedicating part of a year to the study of French. Not really captivated by his study, he apprenticed with his surgeon-father for a year's term. In short order, he tired of the apprenticeship, which lacked intellectual excitement.[5,8,9] He sought better organization and a dedicated commitment to academics. Warren's discontent is fascinating, particularly regarding the practice of his father, a professor and founding surgeon of Harvard Medical School. One has to ponder what was the level of medicine and surgery being practiced in the United States as the nineteenth century dawned.[1-9]

In 1799, at the age of 21, young Warren redirected his academic ambitions from America to the European continent. The Old World offered a great source of enlightenment for the pertinent medical questions of the day. Warren began his European quest by attending Guy's Hospital in London, where the noted professors William and Astley Cooper taught anatomy and surgery. Warren was delighted to hear these especially qualified European masters, who had a unique, elegant presence as they shared their expertise. This education was markedly different, he probably thought, from what he had seen in Boston. Coming to Europe had been the right decision he began to

Reminiscences on Surgery, History and Humanities,
edited by Luis H. Toledo-Pereyra. ©2007 Landes Bioscience.

Figure 3. Portrait of John Collins Warren (1778-1856) at an early stage of his professional career.

believe. Harvard was no match for the London medical schools, he possibly murmured to himself.

In September 1800, Warren transferred to Edinburgh, Scotland, where medicine had blossomed, becoming an example to the world and representing the best of medical science. Warren studied medicine, chemistry, and surgery under John Bell, Charles Bell, James Hope, and Alexander Monroe II.[5,8] Warren's stay in Edinburgh was highly productive and extremely informative. Medicine and surgery in Scotland had no equals elsewhere in the world, he most likely commented to himself.

Warren continued his extraordinary sojourn in Paris, where he absorbed the best of Parisian medicine and surgery for one year. He located the great French professors of the day and registered to attend their famous lectures. George Cuvier and Guillaume Dupuytren were particularly distinguished among his outstanding teachers.[8] Furthermore, Antoine Dubois, Napoleon's surgeon, provided him with excellent surgical instruction at the Hospice of the School of Medicine.[5,8] As great opportunities appeared, Warren took advantage of each one. His rewarding travels and attendance to the best teachers of European medicine and surgery culminated in the best medical education possible at the time (Fig. 3).

In 1802, after three years of the best medical and surgical teachings, Warren prepared to return to his beloved Boston. His father had requested his presence, and soon they ignited a strong and worthwhile partnership, one that lasted for many years until his father's death in 1815. Shortly after, Warren was offered and gratefully accepted the position previously occupied by his father as Hersey Professor of Surgery at Harvard Medical School, an appointment maintained until 1847 when Warren retired as professor emeritus.[5,8,9] Warren committed long hours to his surgical career and medical school position at Harvard.

The contributions of John Collins Warren to the medicine and surgery of colonial Boston proved legendary in terms of surgical practice, medical education, academics, scholarship, the founding of an extraordinary hospital and of a unique and superb medical journal. Here are the specific facts: (1) He performed all possible contemporary surgical procedures, including operations for strangulated hernia, vascular surgery as performed in Europe, cataracts, tumors, diseased bone, and morbid extremities; (2) He was Dean of Harvard Medical School from 1816 to 1819, and at the end of his tenure, he received an honorary MD. In 1802, St. Andrew's University in Scotland had bestowed a similar degree; (3) He published a great number of papers and several books dealing with surgical issues;[5,8,9] (4) He founded (in association with James Jackson, professor of medicine) the Massachusetts General Hospital in 1821, an institution that would set standards for excellent medicine in the years to come; (5) He established in 1828 together with Jackson, John Gorham, Jacob Bigelow, and Walter Channing, the *New England Journal of Medicine and Surgery*, today known as one of the world's top journals under a shortened name.

A special event of considerable consequences occurred in Warren's professional life on October 16, 1846.[8,9] On that day, the first successful inhalation administration of ether occurred. The anesthesia was administered by William Morton to a patient of Warren's who underwent excision of a parotid gland tumor without pain or secondary problems. This previously unimaginable event created extraordinary possibilities for patients and surgeons alike.

Henceforth, surgery could be performed without pain and with the luxury of time in which to excel at the most exquisite details of the surgical technique. In 1848, two years after the introduction of anesthesia by Morton, Warren analyzed his interest in the use of ether,[8] as later quoted by Talbott:[5]

> *"Many years have elapsed since I myself used ethereal inhalation to relieve the distress attending the last stage of pulmonary inflammation. So long ago as the year 1805, it was applied for this purpose, in the case of a gentleman of distinction in the city, very frequently since, and particularly in the year 1812, to a member of my family, who experienced from it great relief, and still lives, to give testimony to its effects. The manner in which it was applied was by moistening a handkerchief and placing it near the face of the patient.*
>
> *A new era has opened to the operating surgeon! His visitations on the most delicate parts are performed, not only without the agonizing screams he has been accustomed to hear, but sometimes with a state of perfect insensibility, and occasionally even with the expression of pleasure on the part of the patient. Who could have imagined that drawing the knife over the delicate skin of the face might produce a sensation of unmixed delight! that the turning and twisting of instruments in the most sensitive bladder might be accompanied by a beautiful dream! that the contorting of anchylosed joints should coexist with a celestial vision! If Ambrose Paré, and Louis, and Desault, and Cheselden, and Hunter, and Cooper, could see what our eyes daily witness, how would they long to come among us, and perform their exploits once more!"*

In 1850, Warren became the president of the recently established American Medical Association. He had spent many hours on behalf of this new organization and was ready to become its spokesman. In addition to his medical and surgical causes, Warren belonged to the Society of Natural History, the Society of Natural Philosophy, the Temperance Society, the Massachusetts Agricultural Society, and played an integral part in the physical education movement. He dedicated himself to advancing the objectives of these selected societies and was eager to contribute to their goals and well-being.

Throughout his accomplished life, John Collins Warren commanded respect and admiration from his peers, colleagues, the medical and surgical community at large, and the patients with whom he came in contact. In short, Warren was a true leader in promoting and advancing the cause of surgery, medical education, and patient care, in addition to all his nonmedical commitments. He will be remembered as a man of conviction and dedication to all his Boston compatriots.

References

1. Warren E. The Life of John Collins Warren, M.D., Compiled Chiefly from His Autobiography and Journals. Boston: Ticknor and Fields, 1860.
2. Traux R. The Doctors Warren of Boston, First Family of Surgery. Boston: Houghton Mifflin, 1968.
3. Arnold HP. Memoir of John Collins Warren. Cambridge, MA: John Wilson and Son, University Press, 1882.
4. Warren IInd JC. To Work in the Vineyard of Surgery. In: Churchill ED, ed. Cambridge: Harvard University Press, 1958.
5. Talbott JH. John Collins Warren (1778-1856) in A Biographical History of Medicine, Excerpts and Essays on the Men and Their Work. New York: Grune and Stratton, 1970.
6. Warren JC. Etherization; With Surgical Remarks. Boston: WD Ticknor, 1848.
7. Warren JC. On an Operation for the Cure of Natural Fissure of the Soft Palate. Amer J Med Sci 1828; 3:1-3.
8. Joseph DG. Warren, John Collins (1778-1856). In: Johnson A, ed. In Dictionary of American Biography. New York: Charles Scribner's Sons, 1964.
9. Gariepy TP. Warren, John Collins (1842-1927). In: Johnson A, ed. In Dictionary of American Biography. New York: Charles Scribner's Sons, 1964.

CHAPTER 18

Valentine Mott

Luis H. Toledo-Pereyra

"The Congress returned to New York's City Hall in 1785, twenty years after the meeting of state delegates in that building. Here, in 1789, George Washington was inaugurated president and the first Congress under the Constitution was convened".[1] This was the state of affairs in the old capital when Valentine Mott (1785-1865) was born in Glen Cove, Long Island, outside of New York City. Samuel D. Gross (1805-1884),[2] the eminent surgical professor and scholar of the second half of nineteenth-century America, highly praised Valentine Mott as the unquestioned American surgical leader of the first half of the nineteenth century.[3,4]

Valentine Mott had a uniquely privileged medical and surgical education. His father, Henry Mott (1757-1840), was a well-respected Quaker physician of Long Island with prominent medical connections in New York, particularly his association with Samuel Bard (1742-1821), an influential medical educator at Columbia Kings College. Henry Mott saw that his son acquired elementary classical teachings at a private seminary. Further studies in secondary education proceeded until 1804 when Valentine Mott began an apprenticeship in the office of his cousin, Valentine Seaman (1770-1817), a distinguished surgeon in New York City.[2-9]

During Valentine Mott's early medical career under the direct tutelage of his cousin, surgery occupied the most important interest of his young mind. While working in practical matters with Dr. Seaman, he attended a complete course of medical school lectures at Columbia Medical College. In 1806, he received his medical doctor degree with special distinction. Particularly remarkable is that his thesis did not show "an active predilection for a future career in surgery," as indicated by his noted medical biographer, surgeon and historian Ira Rutkow.[5] Valentine Mott rather referred to the chemical and medical properties of the *Statice Limonium of Linnaeus*, a drug sometimes used for the treatment of diarrhea,[6] as the central subject of his thesis.

Reminiscences on Surgery, History and Humanities,
edited by Luis H. Toledo-Pereyra. ©2007 Landes Bioscience.

This rapidly changed when Valentine Mott, in 1807, attended surgical studies abroad under the auspices of the most renowned world surgical leaders of the time. For two years, he visited the great surgical centers of London and Edinburgh. His major area of concentration focused in London with the surgical star of the day, Sir Astley Paston Cooper (1768-1841), the great medical personality of Guy's Hospital.[10-12] At the time, "the two best known men in London were George the IV and Sir Astley Cooper".[11] Sir Astley Cooper was a dedicated clinical surgeon, a superb surgical lecturer, a master teacher who offered the best to the needs and intellectual curiosity of the young Valentine Mott. And this was perfect and highly welcomed by the novice from New York.

After six months with the surgical leader at Guy's, Valentine Mott moved within London to study at St. Bartholomew's, London, St. Thomas's, and St. George's Hospitals under the guidance of notable surgeon masters John Abernethy, William Blizzard, Henry Cline, and Everard Home, respectively.[4,5] Great opportunities abounded for Valentine Mott, but perhaps the highlight of his studies in London was the observation of carotid artery ligation for aneurysm by pioneer surgeon Sir Astley Cooper.[7] This operation left an indelible mark on the mind and surgical spirit of the aspiring surgical master.

After a year of intense and arduous postgraduate studies in London, Valentine Mott was not content with his training, and still wanted to visit and review the work of other notable surgical teachers. With this in mind, Valentine Mott left for one of the great medical centers of the day, the city of Edinburgh. In this important surgical center, he studied under John Thomson at the University of Edinburgh and John Bell in private practice. Besides his surgical instruction, Valentine Mott enjoyed the extraordinary cultural aspects of the Scottish world's intellectual center.[5,7,13]

Another year in Edinburgh and surrounding cities was not sufficient for Valentine Mott, who searched for other significant surgical specialty centers in northern Europe, such as Holland and France, where he could enhance his abundant and steadily increasing surgical knowledge. However, war between Great Britain and France prevented him from extending his trip outside of England and Scotland.

In 1809, Valentine Mott concluded his European tour and returned to New York City. America was ready to receive the well-trained and excellently prepared surgical son. Immediately upon his arrival, he set up shop in the city and began a course of private instruction in surgery and anatomy.[5,7] Because of the increased notoriety, Valentine

Mott attracted the academic world, and Wright Post (1766-1828), professor of surgery at Columbia Medical College, invited him in 1810 to become a lecturer and demonstrator of operative surgery at the college. By 1811, Valentine Mott was elected professor of surgery at the medical school at the young age of twenty-six.

In 1813, Columbia Medical School united forces with the College of Physicians and Surgeons. Valentine Mott continued as professor of surgery in the new institution until 1826. That year, because of the resignation of the Columbia faculty, they all moved to establish an independent medical school in the City of New York under the auspices of Rutgers Medical College of New Jersey. Valentine Mott remained in this newly created medical group until 1831 when it was forced to close due to technicalities. From 1831 to 1834, Valentine Mott returned to the Columbia College of Physicians and Surgeons as a professor of operative surgery and surgical and pathological anatomy. His retirement in 1834 was due to deteriorating health.[3,5,13]

In 1835, Valentine Mott began an extended European tour that lasted until 1841. The first order of business was to visit his dear professor and now prominent colleague, Sir Astley Cooper, in Great Britain. They reminisced together and recalled the surgical procedures in which they had mutual interests. Ireland, France, Belgium, Holland, Germany, Italy, Greece, Egypt, Asia Minor, and Turkey were also favored by an extensive visit.[5,7,13] This comprehensive and enlightening trip offered Valentine Mott an opportunity for long-term invigoration. When the offer came to serve as the first chairman and professor of surgery and surgical anatomy, and president of the faculty of the newly established University of the City of New York, he accepted the position at once.

From 1841-1850, Valentine Mott intensely worked at the University of the City of New York (Fig. 4) until his health regressed again to a precarious state. For the third time, he traveled to Europe, where he stayed until the fall of 1851. On his return, he resumed his old position at the University City of New York until 1853, when he fully retired to the position of emeritus professor of surgery with the College of Physicians and Surgeons. Valentine Mott died in 1865.[3-7,13]

Valentine Mott has been very appropriately considered by various medical historians, particularly surgeon Ira Rutkow, as the most deserving physician to be named the father of American vascular surgery.[5] Other names have been put forward as potential candidates, names such as Rudolph Matas,[14] Rene Leriche (a Frenchman),[15] Alexis

Figure 4. Portrait of Valentine Mott (1785-1865) Circa 1846.

Carrel (born in France, but performing much of his work in the United States),[16] and Michael De Bakey. There is no question that all of them are distinguished personalities in their own right. However, Valentine Mott, in a time when no anesthesia was readily available and no antibiotics were present, realized incredible feats in vascular surgery. Because of this, I agree with the position of Rutkow; Mott should be considered and clearly should be given the title of father of American vascular surgery. Others, who came almost one hundred years later, could be considered perhaps as the father of modern American vascular surgery.

No other surgeons in Valentine Mott's time or immediately there-after had performed nearly as many arterial ligations for aneurysms as the record sustained by the extraordinarily talented, New York surgeon. He performed a total of 138 ligations for aneurysms of the major arteries.[5] Of these, there were 57 femoral arterial aneurysms ligated, 51 external carotid arterial aneurysms, ten popliteal arteries, eight subclavian, six external iliac, two common carotid, one common iliac, and one innominate arterial aneurysm.[5-7,13]

As a surgeon, Valentine Mott did not limit himself to the vascular system. He was recognized as a lithotomist as well, performing 165 lithotomies in his lifetime. He was equally busy with bone and joint procedures, performing nearly 1000 amputations throughout his surgical career. Other complex operations of the face, nose, and clavicle entered his repertoire.[3] Valentine Mott delved into any procedure that he felt was feasible, in spite of the lack of anesthesia (introduced in Boston by Morton and Warren in 1846) during most of his active surgical practice. Antisepsis, according to Lister's principles, was not available until 1867. Transfusions and antibiotics were missing from the surgeon's armamentarium.[17] These drawbacks did not stop Professor Mott from realizing his surgical dreams and goals of offering expertise to those in great need. Consequently, he saved many lives.

During the American Civil War (1861-1865), Valentine Mott actively participated in helping those who had been wounded in battle. He visited prisoners of war and was deeply affected by their acute concerns and severe living conditions. He was a dedicated member of the US Sanitary Commission and produced important works in support of the use of anesthetics and the treatment of hemorrhage associated with gunshot wounds.[18,19] His support of the injured was legendary. His commitment to help was without question. His determination to improve the lives of those in pain was unique.

Valentine Mott was not an academic and scholarly writer. He did not contribute to the writing of a surgical textbook. The closest he came was supervising the translation from French of Velpeau's three volume *New Elements of Operative Surgery*.[5] The famed Samuel Gross from Thomas Jefferson University remarked on the lack of interest in writing of the notable New York surgeon: "he could wield his knife much better than his pen…".[5]

Several prominent awards were bestowed on Mott's work and persona. He received an honorary fellowship from the Kings and Queens College of Physicians of Ireland, an honorary MD from the University of Edinburgh, and LLD from the University of the State of New York, and the Valentine Mott professorship of surgery from the College of Physicians and Surgeons of Columbia honored him.[6,9]

As was master surgeon John Hunter (1728-1793) two generations before, Valentine Mott was a dedicated collector of medical specimens. His long stay in London and his association with Sir Astley

Cooper stimulated his interest in collecting anatomical samples. According to medical historian Ira Rutkow, in 1865 Valentine Mott had a collection of more than 1000 specimens.[5] Mrs. Valentine Mott donated this large collection to the City of New York and, unfortunately, a fire at 58 Madison Avenue in 1867 destroyed everything pertaining to this invaluable work of love, study, and sacrifice invested by Valentine Mott. Losing this heritage represented a real personal and public tragedy.

Valentine Mott was a family man in spite of his long working hours and multiple commitments. He married Louisa Dunmore Mums in 1819. They procreated nine children. Except for his long-term and fairly extensive European and Middle Eastern trips, he spent the rest of his time in the United States with his family, friends, and coworkers. I do not have information regarding the whereabouts of family and close relatives except for his son-in-law, William H. Van Buren, who became the first professor of genitourinary surgery in the United States.[7]

How can you define the personality of Valentine Mott? He was a highly committed individual, precise, hardworking, detail oriented, and of serious character.[5,13] Other biographers describe him as respectful, compassionate, and trustworthy.[9] He appeared "courteous, kind, and polished"[5] in his dealings with students and faculty. He was a considerate teacher but sometimes when lecturing he could be "overbearing" and "egotistical".[5]

In short, we can close by saying that Valentine Mott was considered by Professor Samuel Gross "the leading American surgeon of the first half of the nineteenth century".[8,13] Sir Astley Cooper said, "he [Mott] had performed more of the great operations than any man living".[4] Valentine Mott represented, without a doubt, one of the most advanced and courageous surgeon leaders of the time. Together with Zabdiel Boylston (1679-1766),[20] Philip Physick (1768-1837,)[21] Ephraim McDowell (1771-1830),[22] William Beaumont (1785-1853),[23] John Collins Warren (1778-1856), and other early nineteenth century surgeon pioneers, Valentine Mott integrated the highest medical science of the day into his surgical practice.

After a long and productive surgical career, on April 17, 1865, Valentine Mott's life ended. Acute gangrene of the left lower extremity and coronary occlusion were the main causes of his passing. A great era of American surgery was closing. A great surgeon had passed.

References

1. Valentine Mott. Wikipedia, the Free Encyclopedia. (Available at: http://en.wikipedia.org/wiki/Valentine Mott. Accessed December 12, 2005).
2. Annan GL. Valentine Mott, 1785-1865: The Academy's Third President. Bull NY Acad Med 1959; 35:469-471.
3. In: Garraty JA, Carnes MC, eds. Mott, Valentine. American National Biography. New York: Oxford University Press, 1999.
4. Rutkow IM. Valentine Mott and the Beginnings of Vascular Surgery. Arch Surg 2001; 136:1441.
5. Rutkow IM. Valentine Mott (1785-1865), the Father of American Vascular Surgery: A Historical Perspective. Surgery 1979; 85:441-450.
6. Valentine Mott. In: Talbott JH, ed. A Biographical History of Medicine. New York: Grune and Stratton, 1970.
7. Bush RB, Bush IM. Valentine Mott (1785-1865). Investigative Urology 1974; 12:162-164.
8. Batt RE. Mott, Valentine. In: Kaufman M, Galishoff S, Savitt TL, eds. Dictionary of American Medical Biography. Westport, Connecticut: Greenwood Press, 1984.
9. Valentine Mott (1785-1865): Manhattan Surgeon. JAMA 1967; 199:98-99.
10. Sir Astley Paston Cooper. In: Talbott JH, ed. A Biographical History of Medicine. New York: Grune and Stratton, 1970.
11. Hale-White W. Great Doctors of the Nineteenth Century. London: Edward Arnold and Co., 1935.
12. Bettany GT. Eminent Doctors: Their Lives and Their Work. London: John Hogg, 1885.
13. Gross SD. Memoir of Valentine Mott, MD, LLD. New York: D. Appleton and Co., 1868.
14. Ellis T, Widmann WD, Hardy MA. Rudolph Matas: The Father of Vascular Surgery. Curr Surg 2005; 62:606-608.
15. Jarrett F. Rene Leriche (1879-1955). Surgery 1979; 86:736-741.
16. Toledo-Pereyra LH. Classics of Modern Surgery: The Unknown Man of Alexis Carrel-Father of Transplantation. J Invest Surg 2003; 16:243-246.
17. Toledo-Pereyra LH. Vignettes on Surgery, History and Humanities. Austin: Landes Bioscience, 2005.
18. Mott V. Pain and Anesthetics. Military, Medical, and Surgical Essays Prepared for the United States Sanitary Commission 1862-1864. Washington DC: 1865.
19. Mott V. Pain and Anaesthetics. (Available at: http://www.general-anaesthesia.com/misc/pain-anesthetics.html. Accessed December 12, 2005).
20. Toledo-Pereyra LH. Zabdiel Boylston. J Invest Surg; (In press).
21. Toledo-Pereyra LH. Philip Syng Physick: Father of American surgery. J Invest Surg 2003; 16:123-124.
22. Toledo-Pereyra LH. Ephraim McDowell: Father of abdominal surgery. J Invest Surg 2004; 17:237-238.
23. Toledo-Pereyra LH. William Beaumont: First American surgeon scientist and father of gastric physiology. J Invest Surg 2003; 16:55-56.

CHAPTER 19

Samuel D. Gross

Luis H. Toledo-Pereyra

Several years ago, I had the opportunity to write an article in Spanish for a well-known Mexican journal, *Cirujano General*, about the extraordinary life and accomplishments of Samuel David Gross (1805-1884), the quintessential American pioneer who practiced surgery in most of the nineteenth century.[1-23] In the research for this article, I realized again the exceptional character this unique surgeon had, whose incredible story needs to be shared in English. Samuel D. Gross did so much for American surgery that he is considered to be the greatest surgeon of his century. In fact, Isaac Minis Hays,[1] a recognized American surgeon of the time, described Samuel D. Gross as "The Nestor of American Surgery". In the 1884 memoir about Dr. Gross, Isaac Minis Hays summarized the contributions of two individuals in the science and art of surgery in this country.

"In the history of American medicine, in the long list of names of men who have been especially distinguished by their valuable contributions to the science and art of surgery, we find but two who are conspicuously preeminent, and upon whom, each alone in his generation, has been conferred by general assent the honoured title of "The Nestor of American Surgery." Of the two, Philip Syng Physick was gathered unto his fathers nearly half a century ago; and to-day we mourn the loss of Samuel D. Gross. Dr. Physick, although a surgeon of great learning and brilliant capacity, added but little to the literature of our profession, and he left his impress upon American surgery chiefly through the fading tradition of his teachings. But Dr. Gross was, in addition, an industrious and voluminous writer, and has left a precise and imperishable record of his profound study and vast surgical experience".[1]

From farmland origins in eastern Bavaria in the Rhine Palatinate, Gross' ancestors migrated to the American continent in the 18th century prior to the American Revolution. They selected, as one would

Reminiscences on Surgery, History and Humanities,
edited by Luis H. Toledo-Pereyra. ©2007 Landes Bioscience.

expect, Pennsylvania-Dutch country, near Easton, where a dialect of German was spoken in lieu of English. Young Samuel grew in these familiar surroundings where nature, country life, and family values were virtues attained from those times that remained until the last moments of his life.[1-5] This upbringing represented the foundation for Samuel's fertile and productive professional and personal existence.

Throughout his elementary and preparatory schools, he concentrated on excelling in languages, and of these, German and French were his favorites. Later on, he would translate into English many medical works in the most varied specialties. When the time came to decide about a career, medicine was his favorite choice. Since he was six years old, his mother indicated he had expressed interest in the medical field.

Samuel D. Gross had all the necessary attributes to become an excellent doctor. His dedication was without reproach. After a short period of time, less than one year, under the apprenticeship of Dr. Joseph K. Swift, he decided to go back to private schools to enhance his knowledge of the classics, including learning Latin and Greek. Following this experience, he applied in 1826 to Jefferson Medical College where he was readily accepted. At this institution, which was to represent the center of many of his accomplishments, Gross received a special educational experience under the tutelage of the professor of surgery and founder of the college, George McClellan. Under McClellan, Gross learned many of the details of surgical practice and advanced his perception of the surgeon's value and stamina within the medical profession.[1-11]

At the completion of medical school in 1828, Samuel D. Gross went into private practice in the city of Philadelphia.[1] Two years of minimal to no economic success stimulated him to move to Easton where a better practice was more accessible. He enjoyed better professional experiences in Easton where he began other scholarly pursuits on his own, such as cadaver dissections and animal experimentation.[1,11] Samuel Gross was committed to surgical research in theory and practice. He devised and completed several basic research protocols when he was in Easton, later on at Cincinnati, and finally continued in at Louisville.[11] Rabbits, cats, and dogs were utilized in various studies.[1,11] He posed theoretical and practical questions to the research performed in the three places mentioned above. Because of the pioneering surgical investigations initially carried out by Doctor Gross, before anyone else in this country, an appropriate title for him would be "Father of American Surgical Research."

In spite of the difficult economic times affecting Samuel Gross' early professional life, these years were happy with fruitful remembrances. He was free from major responsibilities, he was in control of his time, and he could orient his activities anyway he desired. He represented himself and could move in any possible direction. When he decided to leave Easton in October 1833, his personal and professional fortune had changed. He was considered then the most advanced practitioner of the region.

After five years of general private practice in Philadelphia and nearby communities, Samuel Gross was ready to move into academia. In 1833, he headed west to the Ohio Medical College in Cincinnati where he was the demonstrator of anatomy for two years. From 1835-1840, he became the chair of pathological anatomy in the recently developed Cincinnati Medical College. Samuel Gross threw himself into his new position and promptly searched for avenues of improvement.

In 1840, unexpectedly, Samuel Gross was called to the University of Louisville to occupy the professorship of surgery (and chair) to replace the resigning Joshua B. Flint.[1] This was a unique opportunity for Samuel Gross who had not had a formal experience in surgery and so far in his professional development, he had been oriented towards pathology and general practice. Since there was no formal specialty then, aspiring surgeons would attend surgical cases with physicians who were dedicated to the practice of surgery. Samuel Gross concentrated on learning anatomy, performing animal experiments, and practicing pathological anatomy until the time he would be called to the surgical ranks. The time was 1840, and the University of Louisville was the place. He spent 16 years (1840-1856) at the University of Louisville, except for the few months (1850-1851) he spent as professor and chair of surgery at the University of the City of New York in the place recently vacated by the noted surgical pioneer Valentine Mott.[1,3,8-11]

As time passed in Louisville, Samuel Gross grew accustomed to all details of his life in this charming southern city. He enjoyed the people, his work, his family life, his activities, and everything associated with Louisville. His extensive writings were beginning to give him worldwide recognition which he graciously accepted. The people of Louisville had embraced him as a beloved citizen, and according to Hays, his accurate biographer, "he was regarded as the most prominent surgeon of the southwest".[1] It was not a surprise that he received many offers to teach and practice medicine at other surgical programs such as the University of Pennsylvania. In 1856, he received an offer to return to his medical school *alma mater* Jefferson Medical College.

With the deep respect and gratitude that he had for his old school, he could not turn down the offer. After all, in spite of his attraction and commitment to Louisville, he was returning to his origins. Philadelphia was the last stop of his professional and personal life.

On September 30, 1856, Samuel D. Gross, newly appointed professor and chairman of surgery at Jefferson Medical College, delivered his first clinical lecture at his new place. The new professor was ready to apply his knowledge, experience, and charm at Jefferson. Since the times of McClelland, his first and only tutor of surgery, Gross had dreamed of occupying the seat and position of his beloved master. Alas, the time had arrived!

Immediately upon his arrival in Philadelphia, Samuel Gross establised priorities. In addition to his surgical schedule, he began to work strenuously to complete his monumental treaty on the *System of Surgery*. After many months of dedicated effort toward this grand endeavor, the preface was completed on July 8, 1859.[1,11] This *opus magnus* work contained more than 2,300 pages and an equally great number of illustrations. According to J.C. DaCosta, distinguished and dear disciple of Samuel D. Gross' oldest son, Samuel Weissel Gross, the *System of Surgery* was "the greatest surgical treatise of the day, and probably one of the greatest ever written".[5] This sentiment perfectly describes this extraordinary book. Many surgical principles, pathology, overview of surgical disease, and treatment were outlined so masterfully in Gross' book of surgery. It is particularly important and of great significance that Professor Gross included a description and specific qualifications for a surgeon:[11,22]

> *"Qualifications of a Surgeon—The performance of operations presupposes the possession of certain qualities on the part of the surgeon. It is not every man that can become an operator, even presuming that he has the requisite knowledge of anatomy and of the use of instruments...Courage, like poetry, has often been said to be a gift of nature, and nothing is, perhaps, more true; but it is equally certain that a timid man, by attention to his education, and by constant practice, become, in the end, a good operator...Celsus, long ago, happily defined the qualities which constitute a good operator. He should possess, says the illustrious Roman, a firm and steady hand, a keen eye, and the most unflinching courage, which can disregard alike the sight of blood and the cries of the patient. But the above are not the only qualities, important though they be, which should be possessed by an operator. If he is not honest in his purposes,*

or scrupulously determined, in every case, to cat only with an eye single to the benefit of his patient, and the glory of his profession, he is not worthy of the name which he bears, or fit for the discharge of the solemn duties which he assumes. In a word, such an operator is not to be trusted; for he will be certain, whenever opportunity offers, to employ the knife rather for the temporary 'eclat which may follow its use, than for the good of the individual whom he unnecessarily tortures...Such men, of whom there are, even yet, unfortunately, too many in our profession, deserve the name of knivesmen and knaves rather than surgeons and honest men."

From 1856 to 1882, while Samuel Gross remained the surgeon-in-chief at Jefferson Medical College, his labor was intense and his contributions were exceptional. As in his other posts, he worked in many directions with dedicated vigor. He was a consummated writer, a superb teacher and lecturer, and an exceptional surgeon and innovator. It is impossible to dedicate enough space to his well-known accomplishments in this brief historical review. On this matter, I refer the reader to other works presented in the bibliography that further articulate the achievements of this distiguished surgeon, professor, and researcher[1-23] (Fig. 5).

Samuel Gross was a recognized and skilled surgeon. Operations for hernias, bladder stones, and intestinal obstruction frequently occupied his attention. Furthermore, he introduced new techniques consisting of suturing wounds of the intestines and restoring continuity after the resection of a compromised bowel.[1,3,5,11] He was the inventor of many instruments such as the enterotome, urethrotome, tracheotomy forceps, blood catheter, arterial compressor, tourniquet and apparatus for blood transfusion.[1] Samuel Gross was highly respected by his peers. Samuel Gross was a man of his word, reliable and trustworthy. Isaac Minis Hays admiringly referred again to Doctor Gross in the following way:

"Dr. Gross' majestic form and dignified presence, his broad brow and intelligent eye, his deep, mellow voice, and benignant smile, his genial manner and cordial greeting, remain indelibly impressed upon the memory of all who knew him. He was a man of deep mind and broad views, and he was a model of industry and untiring zeal. He always had some literary work at hand, and he was in the habit of rising early in the morning, generally at six o'clock, and accomplishing

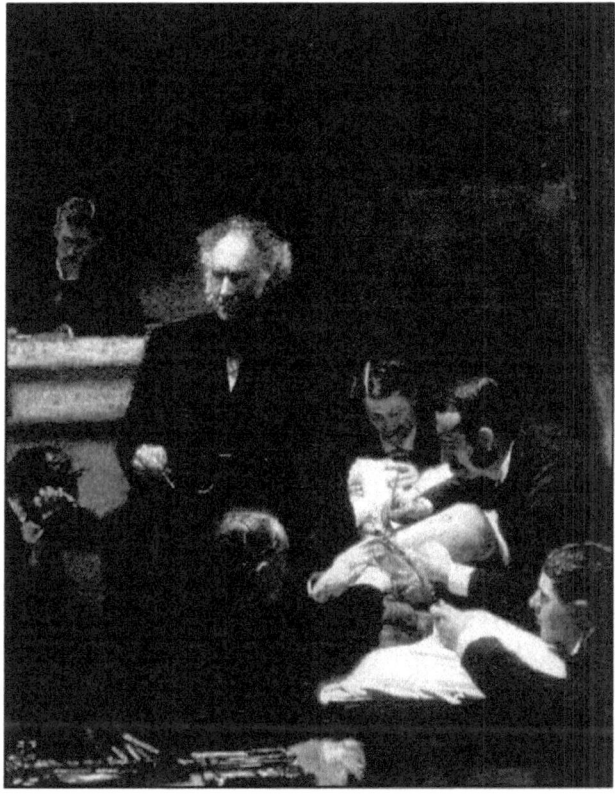

19

Figure 5. The best American painting ever, the "Gross Clinic" (1875) (oil on canvas, 96 x 78 inches) by the superb American artist Thomas Eakins (1844-1916) which clearly reflects the surgery being performed by the noted surgeon Dr. Samuel D. Gross (1805-1884).

> *considerable writing before breakfast. His style was vigorous and pure, and the amount of work he accomplished was simply immense. It is safe to say that no previous medical teacher or author on this continent exercised such a widespread and commanding influence as did Prof. Gross".*[1]

Samuel Gross belonged to a great number of medical societies and founded many of them. He was an established leader of organized medicine. Of particular interest was his commitment to the creation of the American Surgical Association, the American Medical Association, the Philadelphia Pathological Society, and the Philadelphia Academy of Surgery.

Samuel David Gross, pioneer, surgeon, and teacher, was a dedicated author as well. He began in his early years by translating important medical works from the French and German. Gross' interest in writing grew considerably in the following years, and by 1830 his first book on bone injuries appeared.[1] As mentioned previously, in 1839, he published his monumental work on *Elements of Pathological Anatomy*, the first of its kind in English and the one that "made him famous at home and abroad".[1,8-11] This major piece of work solidified his reputation around the world.

In 1851 and 1854, Samuel Gross published two treatises on urological disorders and the management of foreign bodies in the air passages. Both of them represented the definitive works on the matter. In 1859, he put forward one of his most important works, *A System of Surgery, Pathological, Diagnostic, Therapeutic, and Operative*, to which I referred in previous pages. This was the surgical book of the times and the most comprehensive in the United States. In 1861, he produced two books, one on military surgery and another one he edited on *The Lives of Eminent American Physicians and Surgeons of the Nineteenth Century*.[13] Dr. Gross excelled himself not only in the writing of books; in addition, he contributed to many papers and short reports as his biographer, Isaac Minis Hays, so accurately described in great detail.[1]

In 1884, the great surgical professor, researcher, and author died in Philadelphia, the city that witnessed his extraordinary contributions to the field and the unselfishness manner in which he treated his patients and taught his students. One more work that needs to be mentioned is his two-volume *Autobiography*, edited by his sons three years after his death. This book summarized very elegantly the writing career of this giant in American surgery.[17]

It would be fitting to end this brief summation of Samuel D. Gross by using the wise words expressed by Isaac M. Hays at the end of his excellent biographical writing:

> *"In the death of Dr. Gross we have lost one of the brightest examples of the skill and learning, the conscientiousness and assiduity, the patience and perseverance, the dignity and morality by which our profession is truly ennobled. He has left us as a heritage a world-wide reputation which, as we regard it with conscious pride, cannot but stimulate us to a higher sense of our duty to our profession and to our fellows".[1]*

References

1. Hays IM. A Memoir of Samuel D Gross. Am J Med Sci Philadelphia: 1884.
2. Wilson JG. Appleton's Cyclopedia of American Biography. New York: Appleton and Co., 1888.
3. Johnson A, Malone D. Dictionary of American Biography. Vol 4. New York: Charles Scribner's Sons, 1928.
4. Kaufmann M, Galishoff S, Savitt TL. Dictionary of American Medical Biography. Westport: Greenwood Press, 1984.
5. DaCosta JC. Master Surgeons of America: Samuel David Gross. Surg Gynecol Obstet 1922; 35:115.
6. Packard FR. History of Medicine in the United States. Volume I. New York: Hafner Publishing Company, 1963.
7. Brieger GH. Medical America in the Nineteenth Century, Readings from the Literature. Baltimore: Johns Hopkins Press, 1972.
8. Gibbon JH, Samuel D. Gross. Ann Med Hist 1926; 8:136.
9. Rohrer CWG, Professor Samuel D. Gross: America's Foremost Surgeon. Johns Hopkins Hos Bull 1912; 253:83.
10. Wagner FB, Revisit of Samuel D. Gross. Surg Gynecol Obstet 1981; 152:663.
11. Talbott JH, Samuel D. Gross (1805-1884). A Biographical History of Medicine, Excerpts and Essays on the Men and Their Work. New York: Grune and Stratton, 1970.
12. Wagner FB. The Founding Fathers and Centennial History of the Philadelphia Academy of Surgery. Ann Surg 1980; 192:1.
13. Gross SD. Lives of Eminent American Physicians and Surgeons of the Nineteenth Century. Philadelphia: Lindsay and Blakiston, 1861.
14. Wagner Jr FB. Growth and Consolidation of Jefferson Medical College. In Thomas Jefferson University, Tradition and Heritage. Philadelphia: Lea and Febiger, 1989.
15. Cohen AE, The Samuel D. Gross Sesquicentennial. J Ky Med Assoc 1955; 53:981.
16. Garrison FH. An Introduction to the History of Medicine. Philadelphia: WB Saunders Co., 1929.
17. Gross SD. Autobiography with Sketches of His Contemporaries. Philadelphia: George Barrie Publ, 1887.
18. Davis SA, Samuel D. Gross: A Bibliography (1834-1887). Philadelphia: Thomas Jefferson University, 1991.
19. Hall CR. The Rise of Professional Surgery in the United States: 1800-1865. Bull Hist Med 1952; 26:231.
20. Bauer EL. Doctors Made in America. Philadelphia: JB Lippincott Co., 1963.
21. Brieger GH. A Portrait of Surgery: Surgery in America, 1875-1889. Surg Clin North Am 1987; 67:1181.
22. Gross SD. A System of Surgery. Pathological, Diagnostic, Therapeutic, and Operative. 1st ed. Philadelphia: Blanchard and Lea Co., 1859.
23. Toledo-Pereyra LH. Samuel D. Gross: Cirujano, autor y educador. Cir Gen 1994; 16:205.

CHAPTER 20

Daniel Hale Williams

Ralph C. Gordon

The 19th century was difficult for African-Americans seeking a medical education since few schools admitted black students, and the public, including many black patients, had doubts about their capabilities as well. Abraham Flexner in his report on medical education in America in 1910 agreed and stated that "The medical care of the negro race will never be left wholly to negro physicians".[1] This brief study outlines the life and achievements of the man described by his biographer as the "Moses to Negro Medicine".[2] Dr. Daniel Hale Williams was born in Hollidaysburg, Pennsylvania, in 1856, and on the death of his father moved to Baltimore, where he became a barber. He later lived with Harvey Anderson, a barber of modest means in Janesville, Wisconsin, who befriended him as he sought an education.[2,3] Williams attended public schools and the Janesville Academy with Anderson's support and by working as well.

Williams read medicine with Dr. Henry Palmer, who had been Surgeon General of the Wisconsin Regiments during the Civil War. Following that period of apprenticeship, he entered the Chicago Medical School, where he graduated in 1883. This later became the medical department of Northwestern University, and thus he became the first African-American to graduate from that outstanding medical school, and then went into practice in that city. He soon became concerned that opportunities for advanced training were inadequate, since few internships were open to blacks, and opportunities for the education of young black women as nurses were also not available (Fig. 6).

He then embarked on a series of educational and medical ventures that almost single-handedly changed the course of African-American medical history in the United States. The first of these was lobbying in favor of the development of black-run institutions for medical care, postgraduate education of physicians, and nursing education. His 69-bed Provident Hospital in Chicago was the result of his organizing white benefactors and black citizens to

Reminiscences on Surgery, History and Humanities,
edited by Luis H. Toledo-Pereyra. ©2007 Landes Bioscience.

20

Figure 6. Photograph of Daniel Hale Williams.

raise money. It was of particular importance as a hospital approved by the American College of Surgeons, of which Williams was the only black founding member. Provident became the flagship for the black hospital movement.[4] It was noted that some 40 hospitals in 20 states were directly influenced in their development by his institution.[2] He was later called to reorganize the Federal Freedman's Hospital in Washington, DC, which included establishing internships for black physicians, improving the nursing school, and serving on the Howard University faculty in surgery. On his later return to Chicago, he embarked upon one of the most important facets of his work. This was the improvement of surgical education at Meharry Medical College in Nashville, Tennessee, and his presentation of postgraduate surgical clinics for physicians at the Tuskegee Institute in Alabama and at other African-American hospitals. During his first clinic at Meharry, he operated on 25 patients in one week, and his procedures there were attended by both black and white practitioners alike. He continued the teaching clinics there for about two decades, receiving only travel reimbursement, which he donated back to the school.[5,6] These visits added much to the prestige of Meharry and resulted in an increase of enrollment of 60% at the institution. The anesthesia for these procedures was administered by Dr. John A. Andrew of Tuskeegee, who also gave lectures to the local physicians of Nashville. This physician had a strong influence on the development of postgraduate education for physicians of color and helped elevate the status of the African-American physician in the eyes of the profession and the public.

In 1893, Williams performed one of the first cardiothoracic procedures in the United States when he opened the chest of a patient who had been stabbed in the heart. He opened the chest with a wide incision and noted that the cardiac muscle itself was only slightly injured, then sutured the paricardial sac and closed the wound. This procedure had been done only twice previously, and with variable success, but had not been published, so he received considerable notoriety for his work. The patient survived for another 51 years. Williams was also one of the first surgeons to suture a ruptured spleen following trauma, which is now a standard approach, rather than the use of splenectomy. Thus, he did much to advance the cause of the African-American practice of surgery in the United States, proving that blacks could be competent surgeons.[7]

Daniel Hale Williams was also active in the formation of the National Medical Association and served as its first vice-president. This organization was formed in response to the American Medical Association's resistance to having African-American members. It remains an active voice in speaking out for the rights of minority patients and their health care providers. It was formed when a group of black physicians, dentists, and pharmacists met at the Cotton States Exposition in Atlanta in 1895.

In summary, Dr. Williams served as a vital force in American surgical education through the first decades of the 20th century until he died in 1931. He influenced medical education in the South, where he strengthened Meharry Medical College, which along with Howard University graduated most of the black physicians in the United States until the mid-20th century. The extent of his overall contributions to the medical care of African-Americans really cannot be quantified.

References

1. Flexner A. Medical Education in the United States and Canada; A Report to the Carnegie Foundation for the Advancement of Teaching. New York: Carnegie Foundation, 1910:180, 181.
2. Buckler H. Daniel Hale Williams: Negro Surgeon. New York: Pittman Publishing Company, 1968.
3. Williams DH. The Need of Hospitals and Training Schools for Colored People in the South: Detroit. National Hospital Record 1900.
4. Gamble VN. The Black Hospital Movement. New York: Oxford University Press, 1995:1920-1945.
5. Daniel H, Williams MD, LLD. Sims Blue Book and Directory (Illinois). NP: 1924:105.
6. Kenny JA. The Negro in Medicine. Tuskegee, AL: John A. Kenny, 1912:35-37.
7. In: Henderson VJ, Organ Jr CH, eds. Noteworthy Publications by African-American Surgeons. Oakland, CA: Claude H. Organ, 1995:1, 3.

20

CHAPTER 21

John B. Murphy

Ralph C. Gordon

Chicago in the late 19th century was a training ground for many future leaders in surgery in the United States, but singular among them was John Benjamin Murphy, the subject of this historical sketch. He was born on a farm near Appleton, Wisconsin on December 21, 1857 into a stern Catholic family which, though poor, was devoted to bringing out the best in its children. He studied medicine for a period with Dr. H. W. Reilly, his family physician and first role model.[1] With the support of his parsimonious mother he was able to enter Rush Medical College in Chicago, where he graduated in 1879 at age 22. It was there that his independent personality became apparent, as he was not popular as a medical student because of his frequent questions in class, which were interpreted as being self-promoting. On graduation he took the examination for appointment as an intern at Cook County Hospital and scored the highest grade of all the candidates.[2] There he came under the eye of Christian Fenger, surgeon and surgical pathologist who had studied in Europe, who advised him to go there for further study as well.[3,4] Murphy joined Edward W. Lee in the private practice of surgery in 1881. Lee also encouraged his going to Europe and backed this up with an offer to keep his place open in the practice. Murphy then spent 18 months studying in Vienna, Berlin, and Heidelberg under Theodore Billroth and other continental leaders in surgery and then rejoined Dr. Lee, with whom he worked for five years.[2]

His first action that brought notoriety and criticism from his medical colleagues in Chicago occurred in May of 1886 because of a disturbance known as the Haymarket riot. Four workers had been killed in a confrontation between laborers and strikebreakers at The McCormick Harvester factory while agitating for an eight hour work day. The next day a larger labor meeting took place in an area known as Haymarket Square and someone in the crowd threw a bomb which killed seven policemen and wounded 70. The remaining officers then opened fire killing or wounding many bystanders. A call was made

Reminiscences on Surgery, History and Humanities,
edited by Luis H. Toledo-Pereyra. ©2007 Landes Bioscience.

to Murphy's wife and he rushed to the scene of the incident in an effort viewed by many Chicago doctors as "grandstanding" to win publicity for his practice. He operated throughout the night, outlasting the other surgeons present, and won the admiration of the police. Murphy was later called to testify in regard to the deaths of the policemen during the trial of eight anarchists thought to have been involved in the affair. Defense attorneys argued that the police had accidentally killed each other, but he demonstrated the fatal wounds were caused by bomb fragments, and this also began his career as a medico legal expert.[2,5,6] Following the trial, Chicago industrialists supported the building of Fort Sheridan in the northern city suburbs, which was designed to protect them and their property, and it remained an active military installation until recent years.

Murphy enraged his medical colleagues in another episode in 1912 that brought considerable notoriety to his surgical practice and involved president Theodore Roosevelt. The latter was shot in the chest by an assailant in Milwaukee but was not critically ill. The plan was to send him by railroad to Chicago and four outstanding surgeons including John B. Murphy were all called to meet the train at 8:00 AM. The train left Milwaukee early and Murphy met it and its famous passenger at five AM at the Clybourn station. Examination including X-rays revealed the bullet was in the chest wall and had caused no damage to the lung. Roosevelt was soon on his way to continue his campaign activities. Murphy was brought before the Judicial Council of the American Medical Association for possible stealing of patients and other evidence of outlandish behavior but was never disciplined; he later became president of that organization.[2,7]

Christian Fenger had fostered a great interest in Murphy in the use of the microscope, the new science of bacteriology, and the experimental laboratory. He initially had a surgical research laboratory for dogs in a barn behind his house, but his wife secretly had another one built as a gift, and had it fully equipped with the apparatus needed for his work. He took great pride in the laboratory, and in the care he took of his animal subjects, claiming they had as much nursing as his patients, with much of it coming from Mrs. Jeannette Murphy, who also supported him in his practice-building stunts as well.

Murphy's practice boomed and he built an elegant office setting known as the Venetian building which was near Mercy Hospital. He soon moved his surgical practice to that institution after talking the nuns into building a special operating suite for his personal use (Fig. 7). One author has described Murphy as the *Lord of the Knife*

21

Figure 7. A photograph from the Chicago Daily News of Dr. John B. Murphy conducting a clinic at Mercy Hospital, Chicago, Illinois, in 1910.

and in that book gives an extensive pictorial view of his office building and its accouterments.[8] His surgical practice was broad and an analysis of 300 consecutive cases reveals the following distribution: general surgery, 55%; gynecology, 10%; neurosurgery, 5%; orthopedics, 22%; urology, 7%; and other, 1%.[2] He developed several approaches to surgical problems but the most famous were his Murphy button for anastomosis of injured intestines and his advocacy of early operation for appendicitis. Other areas of particular interest included suture of damaged arteries, collapse of lungs for treatment of tuberculosis, gallbladder disease and various orthopedic problems. He was often opposed by the medical community, probably because of his competitive approach to practice, but was generally proven correct in clinical application of his often unorthodox approaches, which tended to fly in the face of the conservative medicine and surgery of the time.[9]

He has often been criticized for his failure to leave disciples, but this is certainly very unfair in that his legacy was a broad one involving many medical students and all physicians who desired to further their education. His frequent clinics at Mercy Hospital were well attended by 100-200 individuals who came to see him operate, discuss the diagnosis of cases, or hear lectures. He served as professor of surgery at the University of Illinois, Rush Medical College, and at Northwestern University where one of his students, Dr. Charles H. Mayo described him as the greatest teacher of clinical surgery through his presentations, while his brother William J. Mayo, graduate of the University of Michigan, characterized Murphy as the surgical genius of their generation. Many photographs taken of him in the amphitheater on close inspection show a smiling face indicating most likely his love of teaching. He often lectured while operating.

Dr. Murphy has another legacy and that is his many contributions to the medical literature of 146 book chapters and journal articles, as well as several hundred case reports (including some duplicates), which appeared in the journal that he started entitled *The Surgical Clinics of John B. Murphy, M.D. at Mercy Hospital, Chicago* or its successors including *Surgical Clinics, Clinics,* and finally *Surgical Clinics of North America,* which has persisted to the present time.[2,10,12] He was also a financial as well as intellectual founder of *Surgery, Gynecology and Obstetrics.*[13]

The next aspects of Murphy's life to be discussed include his participation in the start of the American College of Surgeons, and his presidency of the American Medical Association. He later was elected to the American Surgical Association and to membership in and presidency of the Chicago Medical Society. It should be noted that he had been rejected for membership earlier by these latter organizations, most likely because of his aggressive marketing techniques and possibly personal vendettas against him, since on some occasions the papers he presented at the American Surgical Association went undiscussed at their meetings.

Sir Berkeley Moynihan in a memorial oration given before the American College of Surgeons in 1920, probably best characterized Murphy's persona: "Among those who knew him well he was admired and deeply respected, rather than loved... Except to a very few he was not genial or responsive in friendship... Murphy was beyond question the greatest clinical teacher of his day".[14] His many activities were cut short by his death from heart disease at the relatively young age of 58 at the Grand Hotel on Mackinac Island in Michigan. With his demise the world lost one of its greatest as well as most controversial surgeons.

References

1. Johnson JA. My Years with Dr. John B. Murphy. J Lancet 1964; 84:493-496.
2. Schmitz RL, Oh TT. Murphy's Life. In: Schmitz RL et al, eds. The Remarkable Surgical Practice of John Benjamin Murphy. Urbana: University of Illinois Press, 1993:1-36.
3. McArthur L. Christian Fenger as I knew him. Bulletin, Society of Medical History of Chicago 1923; 3:51-57.
4. Davis L. J. B. Murphy, Stormy Petrel of Surgery. New York: G P Putnam's Sons, 1938:72-73.
5. Stein L, Taft P. Introduction: The Accused and the Accusers. In: Anomyous, ed. New York: Arno and the New York Times, 1969.
6. Avrich P. The Haymarket Tragedy. Princeton: Princeton University Press, 1970.
7. Scafliff HK. Medical Highlights in Chicagoland: Theodore Roosevelt and Dr. J.B. Murphy. Chicago Medicine 1967; 67:397-400.
8. O'Regan SH. Lord of the Knife. J.B. Murphy, Millionaire Surgeon. His Life in Pictures. Amherst: Palmer Publishing, 1986.
9. Isaac G, Hardy MA, Widman WD. John Benjamin Murphy. Current Surgery 2004; 61:439-441.
10. Schmitz RL, Oh TT. The Bibliography of John B. Murphy. In: Schmitz RL et al, eds. The Remarkable Surgical Practice of John Benjamin Murphy. Urbana and Chicago: University of Illinois Press, 1993:167-168.
11. Murphy JB. Murphy's Talks on Surgical and Clinical Diagnosis. Clinics of John B. Murphy at Mercy Hospital. Chicago: 1914:3:403-413.
12. Rutkow IM. A History of the Surgical Clinics of North America. Surg Clin North Am 1987; 67:1217-1239.
13. Martin FH. The Joy of Living. An Autobiography. VI. Doubleday, Doran and Company, 1933:399.
14. Moynihan B, John B. Murphy, Surgeon. Surg Gynecol Obstet 1920; 31:549-573.

CHAPTER 22

Hugh Hampton Young

Luis H. Toledo-Pereyra

When Hugh Hampton Young (1870-1945) began practicing urology in 1897 at the clever recommendation of his praised mentor, William Halsted (1852-1922), and with indirect support from the well-accomplished pathologist and Hopkins founding dean, William Welch (1850-1934), the field was not organized and few well-defined treatments were available for the myriad of human urological diseases.

Young initially had no intention of devoting his best professional years to the study of urinary and genital diseases. He liked general surgery and that discipline stimulated a visit to the Johns Hopkins Hospital. In Baltimore, John Finney (1863-1942), one of the first Halsted trainees, allowed Young to work in the Surgical Dispensary. How this happened is not clear.[1] When a vacancy in the surgical resident staff occurred, he was then considered and accepted for this coveted position. Only professor Halsted could make these critical decisions, which represented the opportunity for a potentially successful surgical career.

Hugh Hampton was a Texas native and had attended the University of Virginia where he received his BA, MA, and MD degrees. After his graduation in 1891, he returned to San Antonio, in his home state, to begin the practice of medicine.[1] He soon realized that he lacked the best medical knowledge of the time. He looked north to Baltimore where great things were happening at Johns Hopkins Hospital and where the most updated medical school was on its way to being incorporated in 1893.[1,2]

Young was a dedicated and resourceful surgical resident. He attended to his duties effectively and promptly. He had a good relationship with his peers and surgeons but was not pleased with Dr. Halsted, who had dedicated no time to advising Young as to his career. This was not unusual at Hopkins, where the professor maintained little or minimal contact with residents other than the chief. The gap was bridged dramatically in October of 1897 when Young

Reminiscences on Surgery, History and Humanities,
edited by Luis H. Toledo-Pereyra. ©2007 Landes Bioscience.

ran into Halstead in one of the long corridors of the hospital.[1] The professor promptly indicated, "I was looking for you, to tell you we want you to take charge of the Department of Genito-Urinary Surgery." Young thanked him and said, "This is a great surprise. I know nothing about genitourinary surgery." "Welch and I said you did not know anything about it, but we believe you could learn," exclaimed the chief surgeon.[1] In this unique way, the incredible career of Hugh Hampton Young was successfully launched.

As common as some genitourinary procedures (such as bladder stone extraction, removal of testicular tumors, treatment of phimosis, and circumcision) were by the end of the nineteenth century, the field of urological surgery had not been fully integrated into medical practice. It was up to the new chief of urology at Hopkins, Professor Young, to create a new specialty, with a new residency program and new operations.[1-5] Extraordinary and unique tasks to be accomplished under any circumstances! Dr. Young set to work with extreme care and dedication to detail (Fig. 8).

The urology residency program was modeled after other Hopkins programs previously well-structured by Halsted in surgery and Osler in medicine. There was no room for miscalculation, so Young carefully analyzed all options and created a premier urological program. He requested seven years of commitment from the accepted residents, so they could fully dedicate themselves to work and breathe urology. Even though Young was a low-key and calm individual, the residency program maintained a high profile and only the best residents were selected.

The first year of residency was spent as a rotating intern in urology, surgery, and gynecology at Johns Hopkins Hospital.[1] The second year was dedicated to pathology under Dr. William MacCallum. Here, the resident learned to recognize all-important urological pathology, from benign findings to severe cancers, a unique and important perspective for the future specialist. In the third year the resident worked at a general surgical service, in particular under the guidance of Dr. T.F. Riggs at Saint Mary Hospital in Pierre, South Dakota.[1] It is not completely clear why Professor Young preferred, if at all possible, that this rotation to be so far from Hopkins. Among the plausible and important reasons were to practice general surgery for one year, be the first assistant to Dr. Riggs, and perform many operations solo. In Halsted's program, these opportunities could not be realized at Hopkins.

Figure 8. Hugh Hampton Young (1870-1945) after a portrait by
Sir William Orpen.

For the fourth year, the urology fellow returned to Hopkins and
continued the urological training, this time by assuming the role of
second assistant resident in urology with an ascending role in assisting in operations, conducting the clinic, teaching, and conducting
research. The following year, the fifth year, the fellow took the role of
first assistant resident in urology, with increased responsibilities in
surgery and teaching and the opportunity to publish his/her own
prior research work. For the sixth year, the fellow went to Ancker
Hospital in Saint Paul, Minnesota, as a complete resident in urology
under Dr. F.E.B. Foley.[1] Dr. Young considered this rotation to be
essential for becoming a urologist, since a great number of cases were
available for residents and Dr. Foley had a reputation as "one of the
most skillful American urologists".[1] In addition, Hopkins could not

accommodate the increased number of residents rotating through the institution.

In the seventh and last year, the fellow returned to Johns Hopkins Hospital as a resident urologist. This time, the fellow was responsible for many of the important functions to be carried out in future practice, namely clinic organization, education, and research. Most importantly, the resident performed all of the operations on patients admitted to the public wards. "I consider it my duty to see that they get as much operative experience as possible," Dr. Young frequently said.[1] It was clear then that the development of the urology residency program at Hopkins represented the best education in any hospital or university in this or any other country. Young and his faculty were particularly proud of such special residency training for future urological surgeons. Others would imitate this superb way of teaching future leaders in urology. Time defined the depth and impact of the program sponsored by Young and Hopkins, and indeed, it came to be considered the premiere program of its day.

In addition to his role as surgical educator, Dr. Young was frequently in the operating room solving important surgical problems.[1-3] In fact, he developed several operations for the management of bladder outlet obstruction. In 1903, he published on the use of conservative perineal prostatectomy for the treatment of benign prostatic disease.[4] Two years later, in 1905, he added a notable and significant publication that dealt with the early diagnosis and radical cure of carcinoma of the prostate.[5] He published 40 cases and introduced a new operation, the radical perineal prostatectomy, which is still being performed today, although it is no longer the most common procedure for treating prostatic cancer.

The potential of the new operation for eliminating all prostatic cancer through the perineal route had monumental positive consequences. For the first time, a radical resection of prostatic cancer could be done safely with significantly low mortality. In fact, during a period of six years, he operated on 128 patients with no mortality at all,[6] an incredible accomplishment considering the state of medical affairs at the time.

Young also introduced other operations, such as the "punch operation," utilized for small prostatic bars and contracture of the prostatic orifice.[7] Years later, the transurethral resection replaced the punch operation of Young. Many more innovative developments occurred in the hands of this esteemed urologist.[1-7] Instruments for diagnosis and treatment of urological diseases were devised by the recognized teacher as well.[2]

Many good surgeons trained under professor Young. They received the best urological knowledge and clinical practice of the time. From 1899-1913, Young referred to numerous trainee assistants in his autobiography.[2] In chronological order, starting in 1899, he had William Huger, Joseph Hume, Hugh Trout, Harry Fowler, John Geraghty, Alex Stevens, John Churchman, Alexander Randall, John Caulk, Harry Plaggemyer, Arthur Cecil, and Oswald Lowsley. Many more residents received excellent urological training under the superb expertise of Hugh Young, who remained the chairman of the department until 1941.

The Brady Urological Institute at Johns Hopkins opened January 21, 1915.[1,2] Through Young's excellent medical care and bedside manner and the extraordinary generosity of James Buchanan Brady, the institute became a reality. Patients, faculty, and residents made this facility the most noted urological institute in the nation, contributing greatly to medicine and patient care.

Hugh Young was not only an excellent surgeon, researcher, and educator, but also committed a good part of his life to helping others. His civic contributions included participation in the aviation committee in Maryland, the Lyric Theatre and Metropolitan Opera Company in Baltimore, and the establishment of the Municipal Hospital for Tuberculosis and the School of Engineering at Hopkins.[1,2] By any standards, he was a dedicated civic servant.

Distinguished urologist Young received innumerable honors and awards. He was president of the American Urological Association (1908), the American Association of Genito-Urinary Surgeons (1910), the Medical and Chirurgical Faculty of the State of Maryland (1912), and the Clinical Society (1925). He was recognized by his peers in medicine and urology by being decorated with the Francis Armory Prize and the Keyes Medal.[1,2] The work of professor Young was highly praised throughout his career. Because of his considerable contributions to urological surgery, the development of the first residency in urology, the publication of his book on the *Practice of Urology* (1926) with David Davis and the founding of the *Journal of Urology* (1917), he should be considered, without reservation, the Father of Modern American Urology.

References

1. James Buchanan Brady Urological Institute. (Available at: http://urology.jhu.edu/about/young.php).
2. Young HH. Hugh Young: A Surgeon's Autobiography. New York: Harcourt Brace and Company, 1940.
3. Young HH, Davis DM. Practice of Urology, 2 Vols. Philadelphia: WB Saunders Co., 1926.
4. Young HH. Conservative perineal prostatectomy. JAMA 1903; 41:999-1009.
5. Young HH. Early diagnosis and radical cure of carcinoma of the prostate. Bull Johns Hopkins Hosp 1905; 16:315-321.
6. Hugh Hampton Young. (Available at: http://www.historiadelamedicina.org/young.html).
7. Young HH. A new procedure (punch operation) for small prostatic bars and contracture of the prostatic orifice. JAMA 1913; 60:253-257.

22

CHAPTER 23

Alfred Blalock

Luis H. Toledo-Pereyra

Nothing was more important for Alfred Blalock (1899-1964) than understanding the mechanisms of complicated surgical diseases, in which he invested his time with unlimited intensity. Early in his career, Blalock dedicated himself to finding explanations for the problems associated with shock. Later he turned to thoracic and cardiovascular diseases.[1-18]

Blalock was born in Culloden, Georgia, and attended high school and college in the state. When young Al decided on a medical career, Johns HopkinsMedical School promised the best possibilities for his inquisitive mind. With the help of professors and friends, and his own enthusiasm, Blalock embarked north to Baltimore, Maryland, where the reputation of Hopkins' new medical school was evident, and its influence spreading.

In 1922, Blalock received his MD from Johns Hopkins Medical School, graduating in the middle of his class. He was not as concerned with scholarly extracurricular activities, such as laboratory and investigative work, as he was with routine work, sports, and social functions. He spent long hours socializing and limited hours in the company of medical books. His average performance in medical school probably precluded him from securing a surgical internship at Hopkins; instead he spent a year studying urology under the general direction of Hugh Hampton Young (1870-1945). He then proceeded to serve another year as assistant resident in general surgery. By now, his clear commitment to clinical work and dedication to medicine had considerably improved, and there was no question of his genuine enthusiasm for scholarly works. In 1924, Blalock published two important studies on the understanding and management of biliary tract disease in the *Journal of the American Medical Association and the Johns Hopkins Hospital Bulletin.*[4] Both works added value to the medical literature of the time. In spite of these developments and his new persona, Blalock's formal surgical training was not going to occur at Hopkins.

Reminiscences on Surgery, History and Humanities,
edited by Luis H. Toledo-Pereyra. ©2007 Landes Bioscience.

With no position forthcoming, Blalock asked for help from Samuel J. Crowe, Chief of Otolaryngological Services at Hopkins, who gave him an unusual appointment as extern in his department from 1924 to mid-1925. The real issue was what to do after this yearlong position ended. Crowe had already communicated with Harvey Cushing at the Peter Bent Brigham Hospital, and Blalock was ready to start work there. Fate intervened when young Blalock received and accepted an offer to do his surgical residency with Barney Brooks at Vanderbilt University Hospital. Tinsley Harrison, his very good friend from medical school and Hopkins days, would be a decisive factor in convincing his aspiring classmate of the importance of taking this academic path. Harrison himself had already accepted a similar position at Vanderbilt in internal medicine. Both friends were bound to Nashville now!

On September 17, 1925, Blalock arrived at Vanderbilt with unfocused goals but the clear desire to excel in surgery. Inactivity was not infrequent in the new program being built by Brooks in Nashville. Blalock grew impatient and bored. Again Harrison convinced his friend to concentrate in the surgical research laboratory, which Blalock found extremely challenging and worthwhile.[4] This opportunity created for Blalock what would be an intense pursuit throughout his surgical career.

In 1928, Blalock finished his surgical residency at Vanderbilt and began his long-term commitment to the science and practice of surgery. He remained active in Nashville, achieving the rank of professor at Vanderbilt (1934-1941). In 1941, his fortune turned north once more to the academically fertile grounds of the young but already reputable university atHopkins in Baltimore. Blalock had meticulously prepared himself to occupy the chair of surgery of any distinguished hospital or medical school in the nation. He had taught medical students, prepared surgical residents, worked in the surgical research laboratory, and attended surgical cases in the operating room. So when the offer came, he was ready to become professor and chair of surgery at Hopkins. And that is the way Blalock arrived!

Blalock's greatest contribution to surgery during his tenure at Vanderbilt came from surgical research. In the laboratory, he continuously advanced the basic understanding of the mechanisms of hypovolemic shock. He introduced new therapies to overcome volume depletion in dogs subjected to bleeding. He recommended the use of plasma or whole blood transfusions to ameliorate the ill effects of hemorrhage. These recommendations were considered critical in

Figure 9. Top left: Photograph of Alfred Blalock with his sister, Georgia, in 1921. Center left: Photograph of Blalock while he was at Vanderbilt, circa 1930. Bottom left: Diagram of tetralogy of Fallot. Right: Photograph of Blalock at Johns Hopkins (date unknown).

saving many lives during World War II (1941-1945). Blalock had received incredible support since 1930 from a talented black surgical technician, Vivien Thomas (1910-1985).[9,10,12] Through the years and under the tutelage of Blalock, Thomas superceded the highest expectations that the professor placed on him. With only a high school education, Thomas learned all the details associated with animal surgical procedures and all the required chemistry laboratory techniques from Blalock. Thomas frequently added his own observations and suggestions. He remained at Johns Hopkins even after Dr. Blalock's retirement (Fig. 9). Johns Hopkins University justly conferred a doctorate of law degree on Thomas in 1976 for the significant advances he made to the science of surgery.[12]

At Hopkins, Blalock concentrated on surgery for blue babies, those with tetralogy of Fallot (consisting of interventricular septal defect, right ventricular hypertrophy, pulmonary stenosis, and dextraposition of the aorta). This congenital cardiac malformation occupied a great part of his time and interest, first in the surgical research laboratory and later in the clinic and the operating room. For this medical and surgical odyssey, Blalock enlisted the help of Helen Taussig (1898-1986), a superb pediatric cardiologist, and of Thomas, surgical technician extraordinaire. This committed team of three special individuals completed for the first time in the laboratory what would be, at the time, the best palliation available for children with tetralogy of Fallot, that is the anastomosis of the subclavian artery-to-pulmonary artery or Blalock-Taussig operation.[10]

On November 29, 1944, after many attempts in the surgical research laboratory, the first patient, 15-month-old Eileen Saxon, frail and massively cyanotic, was carried to the operating room for the first palliative repair of tetralogy of Fallot congenital anomaly. Blalock and Taussig agreed to proceed. Thomas, unscrubbed, stood behind Dr. Blalock throughout the procedure, offering suggestions from time to time, especially during suturing of the vascular anastomosis and handling of the tiny and delicate blood vessels. The operation was a success! Eileen's blue color was gone, "a miracle" in the eyes of her mother. Eileen, unfortunately, died later. In spite of this, the operation had accomplished its goal, to demonstrate its feasibility and to realize that it could be performed safely and successfully.

The spirit of the landmark operation in 1944 was captured by William P. Longmire, first assistant and resident surgeon to Professor Blalock. Longmire later recalled:

> *"I must say my enthusiasm for the idea completely disintegrated when I saw the frail cyanotic infant in the oxygen tent on the east ward of Harriet Lane 4. At that time Blalock spoke briefly with the parents (and indicated again the serious nature of the operation). It seemed to me from the way he greeted them that they had discussed the operation prior to the child's admission to the hospital...At the time of the first operation we lacked all of the modern vascular instruments and really had little but the Professor's determination to carry us through the procedure. The child had extensive collateral vessels full of thick dark blood which I of course, had never seen before. The pulmonary artery was identified with some difficulty and was isolated back into the mediastinum. It was amazing to see the Professor gently but blindly insert a right angle clamp into the mediastinum and after dissecting over his index finger, pull*

*out the innominate artery...Vivien Thomas stood in back of
Blalock and offered a number of helpful suggestions in regard
to the actual technique of the procedure".[8]*

After this first case, many others followed. Blue babies and children of different colors with congenital heart problems flocked to
Johns Hopkins waiting rooms. Taussig would diagnose them, and
Blalock would operate on them. In 1945, Blalock and Taussig published a paper in the *Journal of the American Medical Association* summarizing their experience in the first three cases.[4] Fame and worldwide
recognition came with the success of the operation. They both were
invited to national and international meetings and made clinical presentations all over Europe and throughout the United States. They
both received many well-deserved honors. Blalock alone received nine
honorary degrees, gave a great number of named lectures, and belonged to many important surgical and scientific societies, in particular the National Academy of Sciences and the American Philosophical
Society. Blalock reached the presidency of the American Surgical Association, the American College of Surgeons, the American Association for Thoracic Surgery, the Society of ClinicalSurgery, and the
Society for Vascular Surgery. He shared the Lasker Award (1954) with
Helen Taussig and secured, among others, the Rene Leriche Award
(1949) and the Rudolph Matas Award (1950) for himself.

By the 1950s, Blalock had performed more than 1000 operations
to correct congenital heart defects. During the previous decade, almost anyone who had congenital heart problems looked to Hopkins
in Baltimore as the best option for palliation or treatment of heart
malformations. Blalock and his team left an indelible mark on the
history of surgery and human kind. An incredible journey for Alfred
Blalock from the little town of Culloden!

Blalock trained many surgical residents at Hopkins. A total of 45
surgeons completed their training under his masterful guidance. Everyone received the best from the professor, especially those reaching
the most senior level or chief resident position, as we call it today, the
resident level as it was called then. Other writings refer to the complete list of his distinguished disciples.[4] Many of Blalock's students
reached important positions in academia and the surgical world. Most,
if not all of them, dedicated themselves to the practice of cardiovascular surgery, an emerging discipline in the early 1950s. This cadre
of well-trained individuals contributed immeasurably to the practice
of this new and complicated surgery.

Blalock suffered an undue number of serious health problems throughout his life. From pulmonary tuberculosis, discovered during his early years at Vanderbilt, for which he required a long confinement at Trudeau Sanatorium at Saranac Lake, NewYork, to the development of genitourinary problems (hematuria) for which he underwent a left nephrectomy for a hydronephrotic nonsalvageable kidney, to the recognition of gallstones in the gallbladder treated by cholecystectomy, Blalock had his share of medical difficulties. As if these had not been enough, he also had an appendectomy, bilateral hernia repair at different times, excision of rectal villous adenoma, laminectomy with fusion of L1-L2, several cystoscopies for disuria/hematuria, and other medical/surgical procedures. All in all, Blalock survived his medical ailments well, in as much as he died of metastatic cancer originating from his left ureteral stump at the young age of 65 on September 15, 1964.

Family was important for Blalock. He married Mary Chambers O'Brien on October 27, 1930, and for 28 years they lived together until her death on December 13, 1958. Three children blessed their home, William Rice, Mary Elizabeth, and Alfred Dandy. On November 12, 1959, he married for the second time to Alice Seney Waters, a neighbor he had known for many years at a summer home retreat. Unfortunately, this marriage did not last for too long because of Blalock's untimely death.

Blalock's life was one of dedication and commitment to patient care, teaching, and research. His ability to go to the laboratory and investigate the most profound medical problems was legendary and recognized. His discoveries will remain in the annals of surgery as a laudatory example of tenacity, as he progressed from mediocre student to the heights of surgical innovation.

References

1. Blalock A. Mechanism and treatment of experimental shock. I. Shock following hemorrhage. Arch Surg 1927; 15:762-798.
2. Blalock A. Experimental shock: The cause of the low blood pressure produced by muscle injury. Arch Surg 1930; 20:959-996.
3. Allen JG. Alfred Blalock and our heritage. Arch Surg 1964; 89:929-931.
4. Ravitch MM, ed. The Papers of Alfred Blalock, 2 Vols. Baltimore: The Johns Hopkins Press, 1966.
5. Blalock A. Reminiscence: Shock after thirty-four years. Rev Surg 1964; 21:231-234.
6. Alfred Blalock. (Available at: http://www.mc.vanderbilt.edu/biolib/hc/biopages/ablalock.html. Accessed May 6, 2005).

7. Alfred Blalock Surgical Resident Award. (Available at: http://www.mc.vanderbilt.edu/surgery/ablalock-award.html. Accessed May 6, 2005).

8. Surgeon, Alfred Blalock. (Available at: http://www.medicalarchives.jhmi.edu/blbio.htm. Accessed May 6, 2005).

9. Blue Baby Operation Exhibit. (Available at: http://www.medicalarchives.jhmi.edu/page1.htm. Accessed May 6, 2005).

10. First Operations; Blalock-Taussig Shunt. (Available at: http://www.medicalarchives.jhmi.edu/firstor.htm. Accessed May 6, 2005).

11. Pediatric Cardiologist, Helen B. Taussig. (Available at: http://www.medicalarchives.jhmi.edu/tausbio.htm. Accessed May 6, 2005).

12. Surgical Technician, Vivien T. Thomas. (Available at: http://www.medicalarchives.jhmi.edu/vthomas.htm. Accessed May 6, 2005).

13. Alfred Blalock (www.whonamedit.com). (Available at: http://www.whonamedit.com/doctor.cfm/2036.html. Accessed May 6, 2005).

14. American Experience, Partners of the Heart, Legacy. (Available at: http://www.pbs.org/wgbh/amex/partners/legacy/l colleagues blalock.html. Accessed May 6, 2005).

15. The Lasker Foundation, Former Award Winners, Clinical Medical Research 1954 Obituary. (Available at: http://www.laskerfoundation.org/awards/obits/blalockit/shtml. Accessed May 6, 2005).

16. Alfred Blalock Historical Marker. (Available at: http://www.cviog.uga.edu/Projects/gainfo/gahistmarkers/alfredblalockhistmarker.htm. Accessed May 6, 2005).

17. Tetralogy of Fallot. (Available at: http://en.wikipedia.org/wiki/Tetralogy of Fallot. Accessed May 6, 2005).

18. Longmire Jr WP. Alfred Blalock: Personal Reflections. Pasadena, CA: The Castle Press, 1964.

23

CHAPTER 24

Charles R. Drew

Ralph C. Gordon

Charles R. Drew (1904-1950) was a surgeon whose life was cut short by an automobile accident at age 45, yet during that time he made many continuing contributions to the care of surgical patients, and to the acceptance of African-American medical graduates as surgeons. He was born into a middle class black family in Washington, D.C., and after attending elite Dunbar High School in that city, entered Amherst on an athletic scholarship. Upon graduation he took the position of Director of Athletics and instructor in chemistry and biology at Morgan State University in Baltimore, MD. He was able to save enough money there to enter the McGill University medical school in Montreal where he graduated in 1933. He then completed a rotating internship and a year as resident in internal medicine before returning to Washington where he was appointed an instructor in pathology at Howard University. Drew had also continued his interest in athletics while at McGill and won Canadian awards for hurdle and high and low jump, in addition to maintaining an excellent academic record, for which he was elected to the Alpha Omega Alpha honor society.[1]

His appointment to the Howard faculty in 1935 was a fortuitous one since the Departments of Medicine and Surgery had just received new heads through the support of the General Education Board of the Rockefeller Foundation. The young luminary in surgical research at Yale, Edward Lee Howes who was white, had been named chairman. He took Charles Drew under his wing as a surgical resident, and as was later apparent, his future successor as chair of surgery, since the terms of the Rockefeller grant were that he would be replaced after five years by a black surgeon who had served under him. After one year as a surgical resident he was sent to Columbia-Presbyterian Hospital in New York where he was assigned by Dr. Allen O. Whipple to work under the research supervision of Dr. John Scudder who was beginning fundamental studies in fluid and electrolyte management of surgical patients and the expanding field of therapy with blood

Reminiscences on Surgery, History and Humanities,
edited by Luis H. Toledo-Pereyra. ©2007 Landes Bioscience.

Figure 10. Photograph of Charles Drew (1904-1950) at his microscope, obtained from Howard University, Washington, DC.

products. After six months in the laboratory he entered the surgical residency and completed 18 months of surgical training as well as continuing his research. Dr Whipple was apparently quite taken with the abilities of Charles Drew and encouraged him to complete graduate studies for a doctoral degree in medical sciences, which he received from Columbia University in 1940 for his research and dissertation on blood banking[1] (Fig. 10).

Drew returned to Howard as an assistant professor of surgery, continuing the work that had started with the coordination of manufacture and distribution of plasma to British physicians who were treating military and civilian casualties in the beginning months of World War II. In addition, he became an expert on the Russian literature on blood banking and published several papers which made this information available to American physicians. Although many other investigators deserve credit for their work in research on blood plasma, Drew certainly was responsible for bringing the material from theory to practical application, directing the American Red Cross blood bank program, and opening its first unit in New York City. One of his biographers, Charles W. Wynes, has noted that when Drew was requested to go to Europe to inspect the plasma program, the State Department refused to issue a passport since he was considered "too valuable a citizen to expose himself to the rigors and dangers of the European scene at this time..." On return to Howard he became primarily involved in teaching clinical surgery and was a highly sought after speaker for continuing surgical education nationally.[1] In addition to medical education and research, Drew also spent much of his time and energy trying to improve the lot of African-Americans and fighting racism. During that era the collection of blood from black donors and administration to black or white patients was not allowed. Finally the use of plasma from blacks was allowed but all blood and blood products had to be clearly labeled as to racial origin. This regulation had been foisted upon the Red Cross by order of the U.S. Military and he was powerless to change it. Another major disappointment and a cause of bitterness was his inability to be elected to membership in the American Medical Association. Although there was no national rule against the appointment of black members, and many states had them, the final decision was at the option of the local medical society in the District of Columbia, which did not accept black members, and he fought the decision on several occasions. Similarly, he was not elected to fellowship in the American College of Surgeons, even though he had scored high on the written and oral examinations of the American Board of Surgery, and had an international reputation for his work. Thus the man who was so valuable to his country that the United States Government would not allow him to leave it in war time could not be given membership in the professional organizations for which he was more than qualified. Sadly, the American College of Surgeons elected him to fellowship only after his untimely death. Why he was rejected by that august body is unclear since Daniel Hale Williams, a similarly outstanding black surgeon had been elected to the organization in 1913 and another, Louis T. Wright had been picked in 1935 as well.[1,2]

The passion of Charles Drew for teaching surgery to practicing physicians contributed to his death. He maintained an active life of operating and teaching at Howard University while accepting many invitations for continuing medical education in other cities. He was on a trip to such a meeting in Tuskegee, Alabama in 1950 with three other physicians from Washington, after a night with little rest, when he ran off the road in rural North Carolina, having fallen asleep at the wheel. He suffered a broken neck and massive head and internal injuries and was taken to the local hospital, where he died after telephone consultation with surgical consultants at Duke University Hospital held out no hope. His care in the hospital was judged to be quite acceptable by the doctors who had accompanied him on the trip yet a false rumor arose that he had not been well cared for by his white physicians and that they had refused to give him a blood transfusion. The latter was not true as the hospital had no blood bank. The story was picked up by both black and white oriented newspapers and magazines and later by politicians as well. It had a long life but hopefully has been laid to rest by repeated statements by his wife, other family members, and his surgical associates. The circumstances of his death have been carefully studied by his other biographer, Spencie Love, who has confirmed the appropriate management of his case, and believes that the origin of the story may center around the death of another black patient who was indeed refused admission to the segregated Duke Hospital.[1,3,4] This book provides a very sensitive analysis of racial interactions in the medical setting.

In retrospect, it is hard to say how much Drew could have achieved could he have continued his work as a surgeon, educator, scientist and African-American leader. His memory is preserved in the many schools that have been named after him, including the Charles R. Drew University of Medicine and Science in Los Angeles which has training programs in all areas of health care.[5]

References

1. Wynes CE. Charles Richard Drew: The Man and the Myth. Urbana: University of Illinois Press, 1988.
2. Archives of the American College of Surgeons. Chicago, IL.
3. Craft PP. Charles Drew: Dispelling the Myth. Southern Med J 1992; 85:1236, 1246.
4. Love S. One Blood: The Death and Resurrection of Charles R. Drew. Chapel Hill: University of North Carolina Press, 1996.
5. Charles R. Drew University of Medicine and Science. (Available at http://en.wikipedia.org).

CHAPTER 25

John Charnley

Luis H. Toledo-Pereyra

In the 1960s, the world of orthopedic surgery turned its full attention to the small town of Wrightington, Lancashire, England, as John Charnley (1911-1982) mesmerized his colleagues with advances that made total hip replacement (THR) a real and routine possibility.[1-13] Charnley understood the hip extremely well and worked out the details for better biomaterials and improved biomechanics to establish low-friction arthroplasty. He introduced the use of bone cement (methyl methacrylate) to better seal the implants in the femur and the acetabulum.[1,4,5,9,13] He made joint replacement a practical and reproducible reality, instituting innovations that endure to the present.

Before him, many others had delved into the field of replacing hips, but without permanent success.[11-13] It was Charnley who took an embryonic operation, removed the barriers, and converted total joint replacement into a well-refined specialty. He directed the important business of developing a new and long-lasting hip for the many thousand of patients seeking less pain and better function. He was a true pioneer in creating a new surgical specialty.

Marius Nygaard Smith-Petersen, the noted Massachusetts General Hospital orthopedist, was the star of the 1920s[11-12] with his hip mold arthroplasty, which utilized cups of glass, Bakelite, and early plastic material. All of them thoroughly failed.[4] In 1949, he revised the cups used to include vitallium, a nonreactive metal alloy of great value to orthopedic surgery.[1,4,5] With the use of cups made of vitallium, Smith-Petersen and others, such as his disciple Otto Aufranc, reported good results in 82% of 1000 cases.[14]

Around the early 1950s, the Judet brothers from France, and later Haboush, Urist, and McBride from the United States, and McKee, Farrer, and Ring from England, made pioneering steps in securing the longterm beneficial effects of total hip arthroplasty.[4] According to Leonard Peltier, up to this time, "*the results were not entirely satisfactory because of problems with loosening of the components and wear*

Reminiscences on Surgery, History and Humanities,
edited by Luis H. Toledo-Pereyra. ©2007 Landes Bioscience.

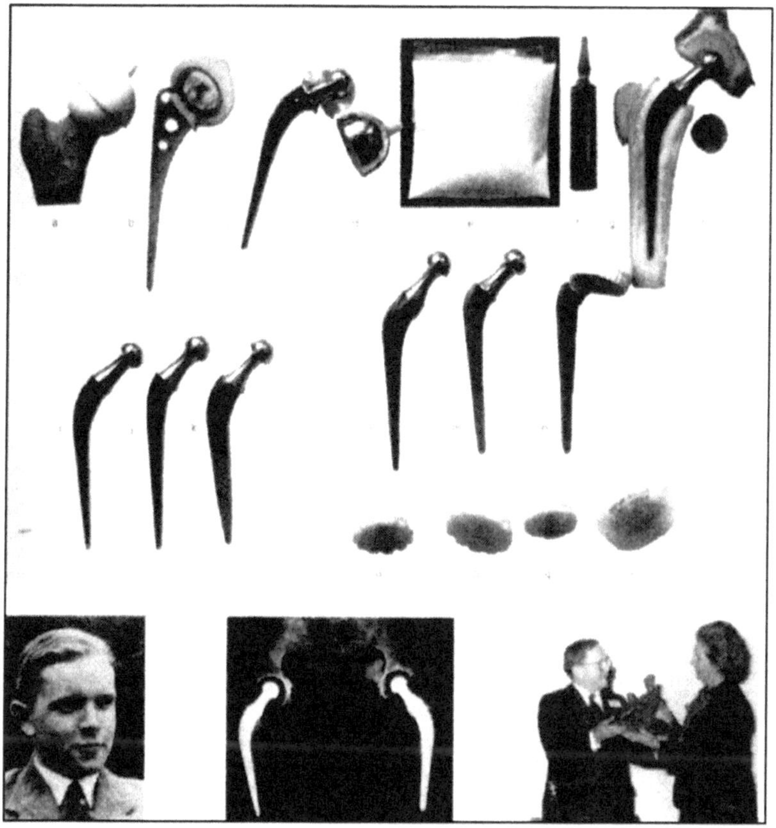

Figure 11. Top: Elements utilized for the Charnley total hip arthro-plasty. Bottom from left to right: John Charnley at the age of 25, radiograph of a patient who had both hips operated on by Charnley 22 and 15 years previously (left and right hip respectively), Charnley receiving the Harding Award from British Prime Minister Margaret Thatcher for his contributions to medical science.

between the opposing metal surfaces. It was John Charnley who led the way in establishing total hip replacement as a useful procedure; one that could be performed by any well-trained orthopedic surgeon, anywhere in the world ".[4] John Charnley completely changed the practice of or-thopedic surgery by carrying the burden of succeeding in a previ-ously failing operation (Fig. 11).

Charnley saw the light for the first time on August 29, 1911, in the small town of Buny, Lancashire, north of Manchester, England. He did all his studies in this area, including his medical schoolwork at the University of Manchester, where he graduated with honors.

He continued his postgraduate orthopedic work at Manchester Royal Infirmary in the late 1930s. His training was interrupted by service in the British Armed Forces in the Middle East. During his military service, he introduced a well-received, innovative way to manage fractures by modifying Thomas' traction method.

At the end of the war, Charnley returned to Manchester Royal Infirmary with renewed interest and dedication. To his clear advantage, he found strong advice and committed support under the tutelage of Harry Platt, a great orthopedic master of the time. In the protective environment of Platt, Charnley advanced his unique and extraordinary ideas about THR. It was the beginning of a life-long love affair with hip replacement surgery.

Pratt was the first and most important fan of Charnley's work, and with his zeal convinced the hospital board to create a special center for surgery and investigation of problems of the hip, utilizing the local hospital of Wrightington in Lancashire for this purpose. Around 1958, Charnley began his dream and significant work at the Centre of Hip Surgery at Wrightington Hospital.[10] As Charnley scaled the ladder of progress and innovation, Wrightington became the primary center for THR in the world. Orthopedic surgeons flocked to the wards and operating rooms of this tiny but great specialty orthopedic hospital to see the professor, the teacher, the extraordinary master, Doctor Charnley work magic with his highly acclaimed procedure.

What then are the specific contributions of Charnley in establishing the principles and practicality of successful THR? We could number the following as his most successful developments: (1) Replacement of the arthritic hip socket with a plastic or Teflon cup and the femoral head with a metal prosthesis of the Moore-Thompson type;[10] (2) Exchange of the Teflon cup for a better high-density polyethylene cup; (3) Use of bone cement, methyl methacrylate, for stabilizing the metal prosthesis or implant in the femur; (4) Development of the concept of low-friction arthroplasty, involving apposition of plastic socket material and the metal head; (5) Use of a smaller, 22 mm-diameter femoral head to reduce contact area and thus friction; and (6) Introduction of clean-air operating techniques to diminish bacteriological contamination during surgery.[1,4,5,9] Many more advances in biomaterials, surface replacement, biomechanics, and the development of new instruments and various steps in the operating technique were the legacy established by this great orthopedic surgeon.

Because of the enormous contributions realized by Charnley during his professional career, he received the highest accolades from his native country and the rest of the world. In the BritishEmpire, Queen Elizabeth knighted him in 1977 for his extraordinary advances in the medical sciences. In the United States, he received the most prestigious award, the Lasker Award for Clinical Medical Research, in 1974, for his notable discoveries in the treatment of patients requiring THR. The only prize that unjustly escaped Charnley was the Nobel Prize. He clearly deserved this award as well. Sir John Charnley was a unique individual whose innovations significantly advanced the surgical care of patients with severe osteoarthritis of the hip. By pioneering safe and long lasting hip replacement, he secured his place as father of modern total joint arthroplasty.

References

1. Coventry MB, Morrey BF. Historical perspectives of hip arthroplasty. In: Morrey BF, ed. Joint Replacement Arthroplasty. 3d ed. Philadelphia: Churchill-Livingstone, 2003:557-565.
2. History of Total Hip Replacement. (Available at: http://www.utahhipandknee.com/history.htm).
3. History of Total Hip Replacement. (Available at: http://thehipdoc.com/history.htm).
4. Peltier LF. A history of hip surgery. In: Callaghan JJ, Rosenberg AG, Rubash HE, eds. The Adult Hip, Vol. I. Philadelphia: Lippincott-Raven Publishers, 1998:3-36.
5. Harkess JW. Arthroplasty of hip. In: Canale ST, ed. Campbell's Operative Orthopedics, Vol. I. 9th ed. St. Louis:Mosby Year Book Inc., 1998:269-313.
6. http://bostontotaljoint.com/thr.html.
7. Charnley J. Anchorage of the femoral head prosthesis to the shaft of the femur. J Bone Joint Surg 1960; 42B:28.
8. Charnley J. Total hip replacement by low-friction arthroplasty. Clin Orthop 1970; 72:7.
9. Charnley J. Evolution of total hip replacement. Ann Chir Gynecol 1982; 71:103.
10. Bernal JJ, Sir John Charnley. (Available at: http://www.biografia/orthopedia.com (Rev. Columbiana de Ortopedia y Traumatologia, Vol. 15, No. 2)).
11. Smith-Petersen MN. Arthroplasty of the hip: A new method. J Bone Joint Surg 1939; 21:269-288.
12. Smith-Petersen MN. Evolution of mold arthroplasty. J Bone Joint Surg 1949; 30B:59-75.
13. Waugh W. John Charnley: The Man and the Hip. London: Springer-Verlag, 1990.
14. Aufranc OE. Constructive hip surgery with a vitallium mold: A report of 1000 cases of arthroplasty of the hip over a 15 year period. J Bone Joint Surg 1957; 39A:237-248.

Section IV.
The Date, the Winners

CHAPTER 26

1954: The Year of the Transplant

Luis H. Toledo-Pereyra and Alexander H. Toledo

Successful kidney transplantation in the twentieth century represented one of the most extraordinary advances in surgical sciences. Initial efforts concentrated on the technical aspects of the procedure, but it soon became very clear that consistent success would only be reached through a multidisciplinary approach. Surgeons, by necessity, paired with nephrologists and immunologists to uncover the problems associated with rejection and medical complications of the transplanted organ and patient. Throughout the whole span of the previous century, many surgeons participated in the surgical development of kidney transplantation.[1-26] This writing emphasizes the historical evolution of living donor kidney transplantation, utilizing Herrick's long-lasting live kidney transplant of 1954 at the Brigham as the pivotal event.

December 23, 1954

By 9:53 a.m., Ronald Herrick's kidney had been removed at Boston's well-respected Peter Bent Brigham Hospital, and it waited in a basin wrapped in a cold, wet towel.[1] The surgical transplant team was directed by the committed and extremely able surgeons Joseph Murray and Hartwell Harrison, who knew perfectly well that the kidney without blood supply could not tolerate separation from the body for too long. Harrison was responsible for the donor side. In another room, Ronald's identical twin brother, Richard, was being prepared to receive the kidney that was being maintained in the basin. Francis Moore, the stern and dedicated chief of surgery, carried the kidney to the next room. Outside, the established and respected nephrologist John Merrill was pacing the corridors awaiting the final verdict. In one hour and 22 minutes, Ronald's kidney was breathing again. Leading the operation on the recipient side, surgeon pioneer Joseph Murray carefully connected all the blood

---

Reminiscences on Surgery, History and Humanities, edited by Luis H. Toledo-Pereyra. ©2007 Landes Bioscience.

hook-ups, and blood was circulating in its regular channels through the kidney by 11:15 a.m.[1] The kidney's color turned pink and urine output steadily increased. The kidney was unmistakably functioning; the miracle was in process! A live kidney, in suspended animation for more than 60 minutes, had returned to life with intense blood circulation and large urinary output.

For the first time in the history of humanity, a healthy monozygotic twin was donating a healthy living organ to his twin brother, who had a severely diseased and irreparable kidney. A year before, in 1953, Michon and his associates[2] had performed a live related kidney transplant with limited success. Two years before Michon, in 1951, Kuss and his group,[3] Dubost and his associates,[4] and Servelle and his team[5] had all pursued kidney transplantation with poor results. But who conceived this kind of operation? How would a donor reach this pinnacle of altruism? Where did this idea originate? These were just a few questions that remained from that special historical moment when science, medicine, and the virtues of humanity all converged. This single uniquely important point in the lives of these generous and clearly unselfish participants would prove to be equally important in the lives of future generations. History had already been conceived; indeed history was in motion! But, for now, one needed to await the results of the operation, an operation sealed by months of solid experimental work, an operation that had been attempted by Kuss[3] and other French surgeons[2,4,5] before, but with no lasting uremic control, an operation that this time could not fail!

The eagerly awaited results revealed definite success. Surgeons and internists continuously checked the urinary output and blood work at Richard Herrick's bedside. They all took turns checking to see that the transplanted kidney was effortlessly functioning without repose, a marvel of science and nature. No evidence of infection or rejection developed. Control of uremia was at hand. The clinical response proved that technically it was possible to transplant a kidney. Now, the challenge was repeating the procedure. Not an optimistic prospect, since patients in need of a kidney transplant rarely had identical twin siblings ready to serve as donors. What then was next?

After December 23, 1954

Surprisingly enough, Murray and his team luckily found other identical twins, one to two every year[1,20,21,27] (Fig. 12). Kidney transplants were performed without difficulty, and with remarkable clinical success. The critical question was how to prevent rejection and

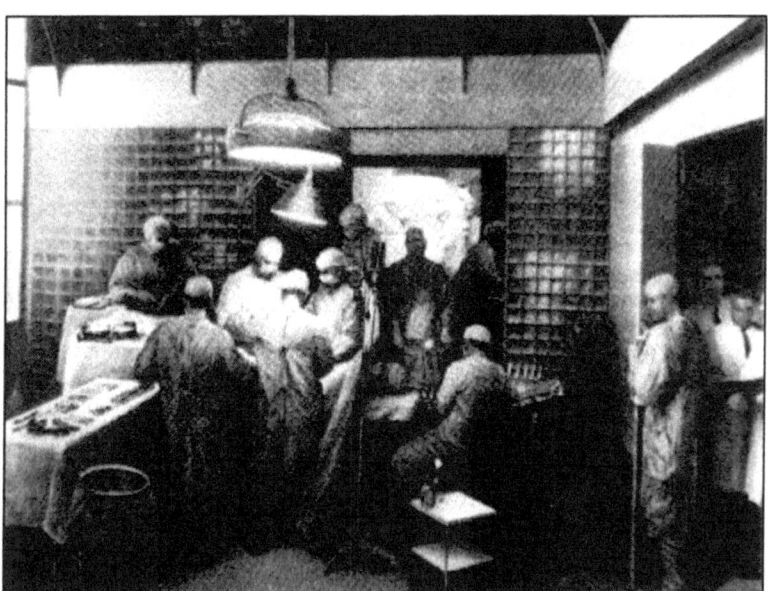

Figure 12. On Friday, December 20, 1996, this extraordinary oil on canvas (70 x 88 inches), "The First Successful Kidney Transplant," painted by Joel Babb in 1995-1996, was unveiled at the Francis A. Countway Library of Medicine in Boston, Massachusetts. As indicated in the booklet printed for the occasion, "Joseph Murray is standing at the patient's right side, to the viewer's left. Harwell Harrison has just completed the [donor operation], entered the hall, pulled down his mask to converse. Francis Moore is seen bringing the precious kidney from the operating room next door. At the hallway door, Drs. Dammin, Thorn, and Merrill have joined Harwell Harrison to discuss the progress of surgery."

obtain some success in patients who were not identical twins.[1,22,28] As a way to overcome genetic dissimilarity, the surgeons began using total body irradiation, learned from the experiments of John Mannick in Cooperstown, New York.[1,10,22,29] Of several transplanted patients, only one, the kidney transplant to John Riteris, was a success.[1] This transplant had been performed with the kidney from his fraternal, not identical twin. In spite of Riteris' encouraging results, he was only one patient, and new avenues, therefore, needed to be intensively pursued.

In 1959, 6-mercaptopurine became an option for kidney transplantation.[25,26,30,31] A year before, Murray and his group had attempted ThioTEPA in rabbits receiving kidney transplants.[32] By 1960, Calne[25] and Zukowski[26] had utilized 6-mercaptopurine with success in dog kidney allografts. About this time, Calne, Alexandre, and Murray were testing at Brigham Surgical Research Laboratories other compounds that had been obtained from Hitchings and Elion of Burroughs Wellcome Laboratories.[31] This time was one of great excitement as the search commenced for new pharmacologic solutions to the overwhelming and thus far unconquerable problem of rejection. Through these studies, an imidazole derivative of 6-mercaptopurine, BW-322, demonstrated the best therapeutic index. The drug was azathioprine, or Imuran, which was going to be utilized for decades to come in the field of clinical organ transplantation. In March 1961, the first unrelated renal transplant patient received Imuran (azathioprine) under the surgical leadership of Murray and his associates. Results were good for a month until the patient died of the drug's toxic effects.[1,22,33-35] Another patient followed with similar results, but in the end, the third patient reached long-term success in April 1962.[22]

Outside Peter Bent Brigham and into the Early 1960s

Encouraged by the superb initial results of Murray and his Brigham team after the 1954 extraordinary kidney transplant success, other surgeons, physicians, and researchers dove into the clinical transplant arena and intensified their efforts. Among these courageous and innovative individuals were Küss, Hamburger and many other French transplant leaders (who had performed the surgical procedure before but with less success refs. 1-8), Calne, Shackman, and Woodruff from the United Kingdom (who had been involved in this field with dedicated interest[25,31,36-38]), and Hume and Goodwin from the United States (who had shown long-term commitment to this endeavor).[18,39,40] These surgeons were established contenders for matching Murray's particularly remarkable results. These uniquely qualified physicians, serious and dedicated professionals, were going to consistently move transplantation forward during the 1960s.[1-10,25,31,36-40]

On March 27, 1962, an important and solid figure in surgical transplantation, Tom Starzl, emerged from the University of Colorado, and with painstaking detail, he began a significant and exceptional series of 83 kidney transplants in humans.[41] All of them were live donor transplants except for three cadaver and six xenograft organ transplants. These transplants were performed in a short period of less than two years. Starzl published his experience in the first extensive book on clinical and practical transplantation in the United States, *Experience in Renal Transplantation*, in 1964.[41] In it, he clearly outlined the extraordinary findings pertaining to the surgical technique, candidate selection, donor evaluation, postoperative care, immunosuppressive therapy, rejection reversal, complication management, and immunological studies with pathological response. This particularly important and significant writing was the fundamental basis for anyone who at the time wanted to organize and develop a new transplant program, in this country and the rest of the world. From 1962 to 1964, the University of Colorado Transplant Program was the most active program in the world, where new and established transplant surgeons would seek guidance as well as attend for sound and practical advice. And advice was plentiful; transplantation appeared to be on the right track!

In the early months of 1963, there were few active programs in the United States. Only Boston (Brigham), Richmond (Medical College of Virginia), and Denver (University of Colorado) were consistently pursuing kidney transplantation.[1,8,22,41] In a matter of months, however, many other institutions began their efforts at establishing kidney transplantation. The example had been set in terms of the specialized surgical technique and the selection of candidates, management of immunosuppression, and in many cases, prevention or treatment of rejection. The most important donor sources for kidney transplantation at that time were live related donors, since cadaver organ transplantation was awaiting the passage of brain death legislation, which began in the United States in 1968 and was completed by 1972. In Europe, similar laws had been incorporated in a similar time frame.

Mid 1960s

By the middle of the 1960s, live related kidney transplantation was fairly well established, and immunosuppression consisted of Imuran and prednisone as popularized by the pioneering efforts of the Colorado chief transplant surgeon, Tom Starzl, and his initial series of 1962-1964.[41] In 1964, at a time when others were searching to improve the immunosuppressive methods, Paul Terasaki, noted tissue-typing expert, used new and advanced histocompatibility testing for the selection of kidney donors from 32 recipients studied from the same Colorado series.[42] Also in 1964, Franksson from Stockholm[43] introduced thoracic duct drainage for eliminating lymphocytes as a unique immunosuppressive alternative in human beings. Two years later, in 1966, Starzl and his innovative Colorado group[44] continued to make significant advances by utilizing for the first time antilymphocyte globulin in clinical renal transplantation with good results. In 1967, John Najarian, an up-and-coming and well-prepared transplant surgeon with a strong scientific immuno-logical background, had recently relocated from California and was diligently revitalizing and reorganizing the existing Minnesota transplant program. Aware of the potential for antilymphocytic prepara-tions, he developed in his laboratories a very effective antilymphoblast globulin with good patient tolerance and acceptable results.[45] In many ways, the immunosuppressive basis for the next two to three decades in transplantation was being established at this point in history. Other developments in patient management produced a more consistent approach to immunosuppression as well.

Late 1960s

By the late 1960s, transplanters appeared to be walking more eas-ily within the clinical field of kidney transplantation. Live related kidney transplantation was a more frequent event. Results were more predictable with one year survival rates of 80%.[46] A consistent tech-nique for surgeons had been developed, the management of live do-nors and recipients was better defined, and a triple method of immunosuppression consisting of Imuran, prednisone, and antilym-phocytic preparations was implemented.[44,45,47,48] Live related kidneys were no longer the only source, as cadaver kidney transplants be-came less rare and more widely accepted. The changes in the practice of transplantation were significant, and the issues began to change as well.[47,49,50] Where previously the knowledge of surgical technique was the most important factor for success, now the management of rejec-tion became of paramount significance instead. Failure became more prevalent in kidneys demonstrating moderate to severe rejection re-sponse, particularly in those patients who showed no reversal after steroid treatment. Live related kidney transplants were less prone to rejection than transplants originating from cadaver donors. As tissue typing became more widespread and precise, immunological causes began to appear as frequent problems in the transplant scenario.

The 1970s and Beyond

The decade of the 1970s introduced to the transplant community a more organized approach. In spite of the extraordinary advances made by the use of recently instituted immunosuppressive regimens, the improved results seen in kidney transplantation were not universal.[9,10,19,24,41,47-50] Some programs reported increased mortality after transplantation.[8-10,24,47-50] Other programs reflected upon the need for better immunosuppression. Causes for higher morbidity and mortality were related in many ways to the appearance of all sorts of infections, for the most part due to enhanced and increased immunosuppression.[9,10,19,24,41,47-50]

Live related kidney transplantation continued to be performed at higher numbers in the 1970s as compared to the previous decade, even though the overall results did not consistently improve throughout this period.[8-10] Dedicated transplant groups persisted in their search for new avenues of improvement under different circumstances. Starzl, in a recent important review paper,[8] astutely labeled this era of the 1970s as the "bleak period" of transplantation, one in which "heavy mortality and devastating morbidity" were the frequent outcomes of many transplant patients.[8]

As the 1970s came to a close, in 1979, Calne and his pioneering group of associates in England[51] published their landmark report on the beneficial effect of cyclosporine A as a single immunosuppressant in cadaver kidney transplantation. Months later, Starzl and his innovative team[52] on the other side of the Atlantic further advanced the effect of cyclosporin A by adding prednisone to the immunosuppression of cadaver kidney transplants, with improved results. The application of this drug to living donor kidney transplantation represented a natural evolution on the use of this new and exciting compound. Other important advances in immunosuppression and novel means to increase organ donation and utilization would be the mark of progress for years to come.

References

1. Murray JM. Surgery of the Soul: Reflections of a Curious Mind. Canton: Science History Publications, 2001.
2. Michon L, Hamburger J, Economos N et al. Une tentative de transplantation renale chez L'homme: Aspect medicolaux et biologiques. La Presse Med 1953; 61:1419.
3. Kuss R, Teinturier J, Milliez P. Quelques essais de greffe du rein chez l'homme. Mem Acad Chir 1951; 77:755.
4. Dubost C, Oeconomos N, Vaysse J et al. Note preliminaire sur l'etude des fonctiones renales de reins greffes chez l'homme. Bull Soc Med Hop Paris 1951; 67:105.
5. Servelle M, Soulie' P, Rougeulle J et al. Greffe d'un rein de supplicie a une malade avec rein unique congenenital, atteinte de nephrite chronique hypertensive azotemique. Bull Soc Med Hop Paris 1951; 67:99.

26

6. Toledo-Pereyra LH, Palma-Vargas JM. Searching for history in transplantation: Early modern attempts at surgical kidney grafting. Transpl Proc 1999; 31:2945-2946.

7. Groth CG. Landmarks in clinical renal transplantation. Surg Gynecol Obstet 1972; 134:323-328.

8. Starzl TE. The mystique of organ transplantation. J Am Coll Surg 2005; 201:160-170.

9. Starzl TE. The Puzzle People: Memoirs of a Transplant Surgeon. Pittsburgh: University of Pittsburgh Press, 1993.

10. Tilney NL. Transplant: From Myth to Reality. New Haven: Yale University Press, 2003.

11. Moore FD. Transplant: The Give and Take of Tissue Transplantation. New York: Simon and Schuster, 1972.

12. Hamilton DN, Reid WA. Yu Yu Voronoy and the first human kidney allograft. Surg Gynecol Obstet 1984; 159:289-294.

13. Carrel A, Guthrie CC. Anastomosis of blood vessels by the patching method and transplantation of the kidney. JAMA 1906; 47:1647-1651.

14. Carrel A. Transplantation in mass of the kidneys. J Exp Med 1908; 10:98-140.

15. Moore FD. A Miracle and A Privilege: Recounting a Half Century of Surgical Advance. Washington, DC: Joseph Henry Press, 1995.

16. Lawler RH, West JW, McNulty PH et al. Homotransplantation of the kidney in the human. JAMA 1950; 144:844.

17. Hamburger J, Vaysse J, Crosnier J et al. Kidney homotransplantations in man. Ann NY Acad Sci 1962; 99:808-820.

18. Hume DM, Merrill JP, Miller BF et al. Experiences with renal homotransplantations in the human: Report of nine cases. J Clin Invest 1955; 34:327-382.

19. Gaston RS, Diethelm AG. Living Donor Kidney Transplantation. In: Gaston RS, Wadstrom J, eds. Living Donor Kidney Transplantation. London: Taylor and Francis, 2005.

20. Murray JE, Merrill JP, Harrison JH. Renal homotransplantations in identical twins. Surg Forum 1955; 6:432.

21. Merrill JP, Murray JE, Harrison JH et al. Successful homotransplantation of the human kidney between identical twins. JAMA 1956; 160:277.

22. Joseph E. Murray-Nobel Lecture. (Available at: http://nobelprize.org/medicine/laureates/1990/murray-lecture.html. Accessed August 11, 2005).

23. The Nobel Prize in Physiology or Medicine 1990-Presentation Speech. (Available at: http://nobelprize.org/medicine/laureates/1990/presentation-speech.html. Accessed August 11, 2005).

24. Terasaki PI. History of Transplantation: Thirty-five Recollections. Los Angeles: UCLA Tissue Typing Laboratory, 1991.

25. Calne RY. The inhibition of renal homograft rejection in dogs by 6 mercaptopurine. Lancet 1960; 1:417.

26. Zukoski C, Lee HM, Hume DM. The prolongation of functional survival of canine renal homografts by 6 mercaptopurine. Surg Forum 1960; 11:470.

27. Murray JE, Merrill JP, Harrison JH. Kidney transplantations between seven pairs of identical twins. Ann Surg 1958; 148:343.

28. Merrill JP, Murray JE, Harrison JH et al. Successful homotransplantation of the kidney between nonidentical twins. New Eng J Med 1960; 262:1251.

29. Murray JE, Merrill JP, Dammin GJ et al. Study of transplantation immunity after total body irradiation: Clinical and experimental investigation. Surgery 1960; 48:272.

30. Schwartz R, Dameshek W. Drug-induced immunological tolerance. Nature 1959; 183:1682.

31. Calne RY, Alexandre GPJ, Murray JE. A study of the effects of drugs in prolonging survival of homologous renal transplants in dogs. Ann NY Acad Sci 1962; 99:743.

32. Porter KA, Murray JE. Homologous marrow transplantation in rabbits after triethylenethiophosphoramide (ThioTEPA). AMA Arch Surg 1958; 76:906.

33. Murray JE, Merrill JP, Dammin GJ et al. Kidney transplantation in modified recipients. Ann Surg 1962; 156:337.

34. Murray JE, Balankura O, Greenburg JB et al. Reversibility of the kidney homograft reaction by retransplantation and drug therapy. Ann NY Acad Sci 1962; 99:768.

35. Murray JE, Merrill JP, Harrison JH et al. Prolonged survival of human-kidney homografts by immunosuppressive drug therapy. New Engl J Med 1963; 268:1315.

36. Schackman R, Dempster WJ, Wrong OM. Kidney homotransplantations in the human. Br J Urol 1963; 35:222.

37. Woodruff MFA, Robson JS, McWhirter R et al. Transplantation of a kidney from a brother to a sister. Br J Urol 1962; 34:3.

38. Woodruff MFA, Robson JS, Nolan B et al. Homotransplantation of kidney in patients treated by preoperative local irradiation and postoperative administration of antimetabolite (Imuran); report of six cases. Lancet 1963; 2:675.

39. Goodwin WE, Mims MM, Kaufman JJ. Human renal transplantation- III: Technical problems encountered in six cases of kidney homotransplantations. Trans Am Assoc Genitourin Surg 1962; 54:116.

40. Goodwin WE, Kaufman JJ, Mims MM et al. Human renal transplantation I: Clinical experiences with six cases of renal transplantation. J Urol 1963; 89:13.

41. Starzl TE. Experience in Renal Transplantation. Philadelphia: WB Saunders Company, 1964.

42. Terasaki PI, Verdevoe DL, Mickey MR et al. Serotyping for homotransplantations—VI: Selection of kidney donors for thirty-two recipients. Ann NY Acad Sci 1966; 129:500.

26

43. Franksson C. Letter to the editor. Lancet 1964; 1:1331.
44. Starzl TE, Marchioro TL, Porter KA et al. The use of heterologous antilymphoid agents in canine renal and liver homotransplantations and in human renal homotransplantation. Surg Gynecol Obstet 1967; 124:301.
45. Najarian JS, Merkel FK, Moore GE et al. Clinical use of antilymphoblast serum. Transpl Proc 1969; 1:460.
46. Murray JE, Barnes BA, Atkinson JC. Fifth report of the human kidney transplant registry. Transplantation 1967; 5:752.
47. Najarian JS, Simmons RL. Transplantation. Philadelphia: Lea and Febiger, 1972.
48. Salvatierra Jr O. Renal transplantation: The Starzl influence. Transpl Proc 1988; 20:343.
49. In: Garovoy MR, Guttman RD, eds. Renal Transplantation. New York: Churchill Livingstone, 1986.
50. In: Toledo-Pereyra LH, ed. Kidney Transplantation. Philadelphia: FA Davis, 1988.
51. Calne RY, Rolles K, White DJG et al. Cyclosporin A initially as the only immunosuppressant in 34 recipients of cadaveric organs; 32 kidneys, 2 pancreas, and 2 livers. Lancet 1979; 2:1033-1036.
52. Starzl TE, Weil IIIrd R, Iwatsuki S et al. The use of cyclosporine A and prednisone in cadaver kidney transplantation. Surg Gynecol Obstet 1980; 151:17-26.

CHAPTER 27

Nobel Laureate Surgeons

Luis H. Toledo-Pereyra

The History behind the Nobel

Nine specially gifted individuals-Kocher, Gullstrand, Carrel, Barany, Banting, Hess, Forssmann, Huggins, and Murray-with surgical backgrounds and/or careers committed to surgery received the unique and coveted Nobel Prize in Physiology or Medicine from 1901 to 2005.[1-55] So which characteristics marked their lives and singular accomplishments?

On November 27, 1895, Alfred Nobel (1833-1896), great inventor and business leader of Swedish origin, gathered his lawyer and witnesses in Paris to assemble the will that was going to change the face of medicine forever.[1] Instructions were given to allocate the interests of his fortune to five areas of human excellence: physiology or medicine, physics, chemistry, literature, and peace. The prize of economics was added in 1968. The amount of each prize in Swedish kroners (SEK) grew from 150,000 in 1901 to 10 million in 2006.[1]

According to Nobel's will, the prize in physiology or medicine is managed and approved by the faculty of the Karolinska Institute in Stockholm. Nominations made by the invited professionals (more than 2500 from around the world) are received by the committee from September to January 31. Then, the plausible candidates are thoroughly evaluated. Candidates can be nominated many times, with due consideration each time. In October, the award is announced after a final vote by the full assembly. According to the archives of the Nobel Foundation, from 1901 to 1951, several great American surgeons were nominated but did not receive the award. Among them, Harvey Cushing, Evarts Graham, and Alfred Blalock were frequently considered.[1]

Since the institution of the first Nobel Award in 1901, the Karolinska Institute, as the Nobel Prize awarding organization, has changed a great deal. In 1901, the professional staff had 19 members

Reminiscences on Surgery, History and Humanities,
edited by Luis H. Toledo-Pereyra. ©2007 Landes Bioscience.

and the Nobel Committee included three members, the president of the Karolinska Institute acting as chairman, a secretary (for 42 years Professor Goran Liljestrand), and another member.[1] In 1970, the faculty represented 61 full professors, with reforms possibly enlarging the number to more than 200. This impeding outlook made it necessary to reorganize the Nobel Committee. In 1977, the Nobel Assembly was established independent of the Karolinska Institute and the Nobel estate. Its entire budget originated form the Nobel Foundation. The Nobel Assembly had, by this time, 50 members, all professors at the Institute. The Nobel Committee was now the executive committee of theNobel Assembly and was integrated by five members and an executive secretary. An ad hoc committee of ten members is added for nine months each year to ensure ample expertise. Complete and detailed information pertaining to the selection of Nobel laureates is presented on the website of the Nobel Foundation (<http://nobelprize.org/>).[1]

"The Greatest Benefit on Mankind"

Alfred Nobel, a discoverer, an engineer, and an industrialist, founded many factories and laboratories throughout the world with the main purpose of enhancing explosives technologies.[1,3] He also produced chemical inventions, such as synthetic rubber, leather, and silk, and accumulated 355 patents by the time of his death in San Remo, Italy, in 1896.[1]

Nobel was particularly interested in stimulating up-scale science, in improving the human mind through literature, and in fostering a good relationship with all the various nationalities in the world. Ironically, in spite of having discovered dynamite, an agent of enormous discord among nations, he believed in peaceful understanding as the main route to solving the problems of society. He worked in many countries and understood the conduct and idiosyncrasies of other peoples. In fact, Nobel was a kind and caring individual, and his legendary will is clear proof of his visionary mind as well as his kind spirit. With his incredible fortune of 31 million SEK at the time of his death, he saw a way to help others in the conquest of their dreams, he saw the need for being a catalyst in improving science and human thinking, he saw a well-defined path to improving society.

It is telling to summarize an important part of Nobel's will here, namely the section explaining his main reason for creating the award for physiology or medicine. According to its founder, the Nobel Prizes are based on conferring "the greatest benefit on mankind"[1]:

"The whole of my remaining realizable estate shall be dealt with in the following way: the capital, invested in safe securities by my executors, shall constitute a fund, the interest on which shall be annually distributed in the form of prizes to those who, during the preceding year, shall have conferred the greatest benefit on mankind. The said interest shall be divided into five equal parts, which shall be apportioned as follows: one part to the person who shall have made the most important discovery or invention within the field of physics; one part to the person who shall have made the most important chemical discovery or improvement; one part to the person who shall have made the most important discovery within the domain of physiology or medicine; one part to the person who shall have produced in the field of literature the most outstanding work in an ideal direction; and one part to the person who shall have done the most or the best work for fraternity between nations, for the abolition or reduction of standing armies and for the holding and promotion of peace congresses. The prizes for physics and chemistry shall be awarded by the Swedish Academy of Sciences; that for physiology or medical works by the Karolinska Institute in Stockholm; that for literature by the Academy in Stockholm, and that for champions of peace by a committee of five persons to be elected by the Norwegian Storting. It is my express wish that in awarding the prizes no consideration be given to the nationality of the candidates, but that the most worthy shall receive the prize, whether he be Scandinavian or not."

Nine Nobel Laureate Surgeons

Nine eminent surgeons of various ages, nationalities, and geographical regions received the Nobel Award at different times during the past century (Table 1). From 1901 to 2000, the Nobel Prize in Physiology or Medicine was awarded to 172 individuals, of whom nine were surgeons (5%).[1] The surgical profession, then, has not been over-represented. Let us review some essential details about these surgeon-awardees (Table 1).[2,3]

It is evident that surgeons occupied the attention of the Nobel Committee in the early stages of this pristine award. From 1909 to 1914, there were four surgeons who reached this maximal accolade. After that, except for the 1923 award for the discovery of insulin,

Table 1. Characteristics of Nobel laureate surgeons

Laureate	Year	Age (yrs)	Country of Origin	Area of Award	Surgical Specialty	Years from Discovery to Award
1. Theodore Kocher	1909	68	Switzerland	Thyroid Surgery and Physiology	General Surgery	17
2. Alvar Gullstrand	1911	49	Sweden	Dioptrics of the Eye	Ophthalmology	3
3. Alexis Carrel	1912	39	France/USA	Vascular Suture and Organ Transplant	Surgical Research	10
4. Robert Barany	1914	38	Austria	Vestibular System	ENT	8
5. Fred Banting	1923	32	Canada	Insulin Discovery	Orthopedics	1
6. Walter Hess	1949	68	Switzerland	Midbrain Function	Ophthalmology	11
7. Theodor Forssmann	1956	52	Germany	Cardiac Catheter	General Surgery/ Urology	27
8. Charles Huggins	1966	65	Canada/USA	Hormones/Cancer	Urology	25
9. Joseph Murray	1990	71	USA	Organ Transplantation	Transplantation/ Plastic Surgery	35

27

Nobel Prizes were rarely given to surgeons.[1-3] Six Nobel surgeons came from Europe (one prize was shared with an American), and the other three from North America—two from the United States and one from Canada.[2,3] The surgical specialties of the awardees offer some interesting fields of concentration: two general surgeons, two ophthalmologists, two urologists (one shared with a general surgeon), and one each in ENT, orthopedics, transplantation, and surgical research. A distinguished group of surgeons all around.

It has been said by many physicians that it takes ten years or more from the time of a discovery to receiving this very special prize. Indeed, in this group of distinguished surgeons, it took an average of 15 years to reach this goal. However, there was significant variability, from one year in the case of Banting to 35 years for Murray.[1-3] Many uncontrollable factors influenced this outcome.

One could categorize the awards based on whether they were applicable to clinical surgery or not (Table 2). It is quite remarkable to see that only three awards, those of Kocher, Carrel, and Murray, had a direct use in the field of surgery, whereas the awards for Gullstrand, Barany, Banting, and Hess had no direct surgical application. Forssmann's and Huggins' awards had indirect surgical use.

Now, let us determine if we can encounter some special virtues or well-defined qualities in this especially select group of Nobel Laureate surgeons. What makes these surgeons so special that they deserve a Nobel Prize? What did they discover or popularize that was so unique? What is it that we need to teach a new generation of American students? I think one word summarizes it all: *innovation!* Only with this innate and/or acquired quality can we advance the field of surgery to levels never seen before. Only with this unique feature can we consistently move ahead. Only with this extraordinary characteristic can we mold the future. Innovation represents everything in the

27

Table 2. Classification of Nobel awards in surgery

Direct Application to Surgery	Indirect Application to Surgery	No Application to Surgery
1.Kocher*	1. Forssmann	1. Gullstrand*
2. Carrel	2. Huggins*	2. Barany
	3. Murray	3. Banting
	4. Hess	

*Practicing clinical surgeons all their lives in the field of the award.

ability of the surgeon to be especially inimitable in his/her opportunity to contribute in an unparalleled manner to society. This is, in a nutshell, what Nobel Laureate surgeons are made of! Even though innovation is a significant and unrivaled quality, by itself it cannot always reach the pinnacles of achievement. Dedicated commitment, keen determination, and defined focus are, in addition to innovation, the required ingredients for a successful and worthwhile endeavor. Nobel surgeons had all these distinguished features, as encapsulated in the letters CDFI, which stand for commitment, determination, focus, and innovation. The integration of these four qualities represents very clearly the world of Nobel Laureate surgeons, the world of real surgical innovators, as well as the world of future surgical leaders (Fig. 13).

Differences among Equals

As one would expect, many differences in their lives and work were evident among the nine Nobel surgeons. For **Theodore Kocher** (1841-1917), the Nobel Award in Physiology or Medicine of 1909 was totally based on his clinical work as a practicing surgeon dealing with problems of the thyroid gland.[4,7-14] Other surgeons, such as Theodor Billroth (1829-1894), had undertaken similar operations but none with the degree of attention that Kocher placed on his surgical technique, particularly respecting the anatomical structures and operating with care on the tissue in and about the thyroid gland. He was a master technical surgeon who performed or oversaw more than 7000 thyroid operations with a mortality incidence of less than 0.2% in 1898.[2,4,7-10] Because of this work, the Nobel Committee considered him one of the few clinicians of the time who deserved such special recognition. Before him, eight individuals had won the Nobel Prize beginning in 1901, and of these, only one, Niels Finsen (1860-1904) in 1903 had practiced pure clinical medicine, specifically the treatment of lupus vulgaris. The other awards were associated with laboratory medicine, physiology, histology, or tropical medicine. Kocher was nominated five times before obtaining the Nobel Award and was of special consideration for the nomination by Vinzenz Czerny, disciple of Billroth and a famous German surgeon of the time. When there was the opportunity for Kocher to nominate in 1913 and 1917, he introduced the name of Czerny twice and the names of Horsley, Cushing, and Halsted once. Cushing was nominated a total of six times from 1917 to 1931, and it is difficult to believe he never won. Halsted was nominated only once by Kocher.[1]

Figure 13. Front cover of the book *Maestros Nobel de la Cirugia*. Please note the portraits of the nine Nobel Laureate surgeons, starting chronologically from Theodor Kocher from the upper left and ending with Charles Huggins on the upper right, with Joseph Murray receiving his 1990 Nobel Award from King Carl Gustaf XVI in the center.

The second Nobel Laureate surgeon, **Allvar Gullstrand** (1862-1930), received his award in physiology or medicine in 1911 for his work on the dioptrics of the eye. Gullstrand was an extraordinary physicist and ophthalmologist who, according to his Nobel biography, taught himself geometric and physiological optics.[1,14-17] His works were of incredible insight, admired by many in the field. Although his particular award related more to the physical sciences, his introduction of focal illumination by means of the slit lamp (1911) vividly caught the attention of practical ophthalmologists. The prominence of Gullstrand in the world of physics was such that he was named to the Nobel Physics Committee of the Swedish Academy of Sciences from 1911 to 1929 and was chairman from 1922 to 1929. When his personal Nobel Award came up for consideration in 1911 in the area of physics, he refused the nomination until he could be considered in the area of physiology or medicine, which he won the same year.[1,2,15-17] As chairman of the Physics Committee, he rejected the work of Albert Einstein as not important enough to receive the Nobel Award. Gullstrand was nominated three times for the Nobel Prize in Physiology or Medicine from 1909 to 1911 until the prize was finally awarded in 1911.[1,2,15-17]

The third Nobel Laureate surgeon, **Alexis Carrel** (1873-1944), received his award in 1912 for work on vascular sutures and the transplantation of blood vessels and organs.[1,19,21-24] Carrel's work was performed entirely in the surgical research laboratory; he never practiced clinical surgery. He began his unique experimental work at Lyons Hospital in 1902 and continued in the Department of Physiology of the University of Chicago from 1904 to 1905 under Professor G. Stewart's support and with the assistance of Charles Guthrie. In 1906, he moved to the Rockefeller Institute for Medical Research in New York where he remained the rest of his professional career.[17,23,24] Work done at these three institutions formed the basis for his prestigious Nobel Award.[1] Carrel was nominated once before in 1912 by Professor Bouchard of the University of Paris, interestingly enough because of his work on the continuation of life in isolated tissues. Carrel, a controversial figure, had a difficult personality and was not easy to get along with. Throughout his professional life, he nominated 19 basic scientists and laboratory physicians for the Nobel Prize. Harvey Cushing was the single clinical exception; Carrell supported Cushing in 1932 for his work on neurosurgery and the physiology of the pituitary gland.[1]

The fourth Nobel Laureate surgeon, **Robert Barany** (1876-1936), received his award in 1914 for his work on the physiology and pathology of the vestibular apparatus.[1-3,25-28] Barany concentrated his clinical practice on otological diseases after being the pupil of surgeon Professor Gussenbauer at the University of Vienna. In 1903, Barany was the demonstrator at the Otological Clinic of Professor Politzer where he made significant contributions. In 1905, Barany published the landmark paper on caloric nystagmus. Other important works followed, addressing the vestibular reaction movements associated with equilibrium and existing nystagmus. His work did not have direct surgical application but formed the basis for clinical examination of problems of the ear. When the Nobel Prize was announced in 1914, Barany was a prisoner of war in Russia, and Prince Carl of Sweden intervened on behalf of the Red Cross for his release in 1916.[1,2,24,25] Barany grew discontent with Austrian scientists who did not appreciate his work. He moved to Uppsala where he died in 1936. Barany was nominated seven times before he obtained the Nobel Award. He had the opportunity of nominating seven people after 1917, four of whom favored Freud as a recipient for his work on the spiritual life and other related fields.[1]

The fifth Nobel Laureate surgeon, **Frederick Banting** (1891-1941), received his award in 1923 for his work on the discovery of insulin.[1-3,29-36] This award was shared with John Macleod, director of the Physiological Laboratory at the University of Toronto. Critical commentaries reached the Nobel Committee because Charles Best did not appear in the winning group. After time passed, it became obvious that the award was given justly and for sufficient reason. Fred Banting did not practice surgery, particularly orthopedics, for most of his career. He was an orthopedic resident surgeon at the Hospital for Sick Children in Toronto from 1919 to 1920. The following year he secured a part-time teaching position in orthopedics at the University of Western Ontario in London, Ontario, Canada. Besides a short stint in general practice, Banting instead consumed himself in studying the problems associated with diabetes.[1-3,31] The early publications originating from the laboratories of von Mering and Minkowski dealing with the use of pancreatic extracts stimulated his curiosity. At the end of 1920, he reviewed with enthusiasm a paper by Moses Baron from the University of Minnesota, outlining the degenerative changes seen in the acinar cells after pancreatic duct ligation, preserving in this way the islet cells and their endocrine function. After visiting with Macleod, the experiments began on April 14,1921.[13] The Nobel Prize was given in 1923, one of the fastest turnarounds from the time of discovery to Nobel fame. Banting was nominated three times the same year he received the Nobel, an extraordinary feat for an extraordinary discovery!

The sixth Nobel Laureate surgeon, **Walter Hess** (1881-1973), received his award in 1949 for his work on the discoveries of the function and localization of the midbrain.[1-3,37-40] He received this award jointly with Egas Moniz (1874-1965), the Portuguese neurologist who introduced cerebral angiography and prefrontal leucotomy.[1] Hess graduated from the University of Zurich where he was assistant in surgery and ophthalmology, and a practicing ophthalmologist. In 1912, he made a critical decision to abandon a lucrative practice in ophthalmology to begin a career in physiology. Following a period of intense training with Professors Gaule and Verworn in Bonn, in 1917, he was named Director of the Physiological Institute in Zurich.[1-3,37-40] The interests of Hess focused on the vegetative nervous system in the dienchephalon as well as in the control of the forebrain. These studies culminated in the consolidation of the desirable Nobel Prize. Hess was nominated 26 times from 1933 to 1949, in a period of 16 years. He lived a long life of productive achievements in two professions, ophthalmology and physiology. He excelled in both fields of medicine in spite of the great differences between them.[1-3,37-40]

The seventh Nobel Laureate surgeon, **Werner Forssmann** (1904-1979), received his award in Physiology or Medicine in 1956 for work on the discoveries concerning heart catheterization and pathological changes in the circulatory system.[1-3,5,41-44] He shared the Nobel Award with two American cardiologists, Andre Cournand (1895-1988) and Dickinson Richards (1895-1973) from Columbia University at New York City. Forssmann graduated with an MD from the University of Berlin in 1929. He received instruction in surgery the same year at the August Victoria Home at Eberswalde near Berlin.[1,2] It was in this hospital that he first attempted heart catheterization in a patient he knew very well, himself. He inserted a cannula of 65 cm through his antecubital vein and walked to the radiology department to determine its location, which was in his right auricle.[5,44] This was not the first and only time that Forssmann used himself to probe his research ideas. On other occasions, he utilized his own body as a research testing ground.[2,5,44] Many in the medical community did not agree with this habit and repudiated his personal experiments. Forssmann practiced in many highly academic institutions such as Berlin and Mainz hospitals. He finished his urology training under Dr. Heysch in Berlin. He was a Sanitary Officer in the Second World War and became a Surgeon-Major before becoming a prisoner of war. In 1945 he was released.[2,5,44] There is no knowledge of how many times he was nominated, because the Nobel Foundation stipulates that this information must be secret until 50 years after the award has been made.

The eighth Nobel Laureate surgeon, *Charles Huggins* (1901-1997), received his award in 1966 for his work on the hormonal dependence of cancer cells.[1-3,45-49] Peyton Rous (1879-1970), a Rockefeller Institute researcher, shared the Nobel with Huggins because of his work on viruses and cancer. Huggins came from Canada to Harvard University where he obtained his MD in 1924. He moved to the University of Michigan for his surgery training and then to the University of Chicago where he remained the rest of his life. He maintained a combined active career in urological surgery and oncological research in the basic sciences laboratories.[45] He was extremely interested in innovation and the opportunities for discovery. He would constantly ask his students, "What did you discover today?" He was also a continual source of stimuli for both young and seasoned researchers in his laboratories. He felt that "the business of finding out was one of the supreme joys of mankind".[45] The business of discovery was the essence of research in medicine, he argued.[45] As in the case of Forssmann, we do not know the number of times Huggins was nominated because of the 50-year rule before disclosure.[1]

The ninth Nobel Laureate surgeon, *Joseph Murray* (b. 1919), received his award in 1990 for work on organ and cell transplantation in the treatment of human disease.[1,6,50-57] Donnall Thomas (b. 1920) shared the Nobel for his work on bone marrow transplantation. Murray attended the College of the Holy Cross near Boston where he concentrated on the humanities. Later, while at medical school at Harvard, he would emphasize the sciences. Medical school was what he "dreamed it would be," particularly his time off from class, which showed him a "life that was rich and full".[6] After a surgical internship at Brigham Hospital, he began his military service at Valley Forge General Hospital in Pennsylvania under the Chief of Plastic Surgery, Colonel James Barrett Brown. Murray learned the immunology of skin grafts and observed that rejection did not occur on identical twins and, therefore, grafts will tolerate their transplantation if no foreign antigens are present. These observations were extremely helpful when planning kidney transplantation. In November 1947, Murray returned from the military service to Brigham Hospital Department of Surgery to continue his surgical residency until 1950-1951 when he completed a six-month stint in plastic surgery.[6] Dr. Francis Moore, Brigham Chief of Surgery, strongly supported Murray and put him in touch with fate and an incredible future when by assigned him to the surgery service supporting kidney patients in need of surgical treatment. In 1954, a case arrived to the office of Dr. Merrill, Chief of Nephrology. The patient had kidney failure and a healthy identical twin brother. Merrill communicated with Murray, and planning began for the possible kidney transplant.[1,2,6] And, as frequently mentioned, the rest is history. According to Murray, three factors came together for the realization of the successful kidney transplant surgery: the knowledge of skin grafting in twins, the development of the artificial kidney for acute dialysis (used first in 1948 in humans) at the Brigham as introduced by Kolff and Walter, and surgical determination based on a strong laboratory background using appropriate techniques for surgical implantation of kidneys in dogs.[1,2,6] With every one of these factors already defined and well-studied, the kidney transplant operation of December 23, 1954, in identical twins was an unquestionable success.[58] Other similar operations followed and a new field of transplantation arrived. As with the two previous Nobel surgeons, we do not know yet how often Murray was nominated.

Conclusion

The great lives and discoveries of Nobel Laureate surgeons should serve as a clear example of how important answers to critical clinical questions can be found, how dreams can be effectively reached, and how imagination can be fully applied to the benefit of mankind.

References

1. Nobelprize.org. (Accessed May 23, 2006. Available at: http://nobelprize.org).
2. Toledo-Pereyra LH, Martinez-Mier G. Maestros Nobel de la Cirugia. Mexico: JGH Editors, 2000.
3. Cosimi AB. Surgeons and the Nobel Prize. Arch Surg 2006; 141:340-348.
4. Isepponi O, Huwiler V, Boschung V. Theodor Kocher's surgical and clinical case presentations. Bull Hist Med 2004; 78:192-194.
5. Forssmann WTO. Experiments on Myself: Memoirs of a Surgeon in Germany. New York: St. Martin's Press, 1974.
6. Murray JE. Surgery of the Soul: Reflections of a Curious Career. Canton, MA: Science History Publications, 2001.
7. Bonjour E. Theodore Kocher. Bern: Paul Haupt, 1950.

8. Weise ER, Gilbert JE. Theodore Kocher. Ann Med Hist 1931; 3:521-529.
9. Jain KM, Swan KG, Casey KG. Nobel prize winners in surgery. Part 1. Am Surg 1981; 47:195-200.
10. Kocher T. Text-book of Operative Surgery. HJ Stiles (trans.). London: A and C Black, 1895.
11. Kocher T. Textbook of Operative Surgery. 5th ed. New York: MacMillan, 1911.
12. Nobel Foundation. Nobel Lectures, Physiology or Medicine, 1901-1920. New York: Elsevier, 1967.
13. Martínez MG, Toledo-Pereyra LH. Emil Theodore Kocher: Cirujano, Maestro y Nobel. Cir Ciruj 1999; 67:226-232.
14. Stevenson LG. Nobel Prize Winners in Medicine and Physiology. New York: Henry Schuman, 1953.
15. Ravin JG. Gullstrand, Einstein, and the Nobel prize. Arch Ophthalmol 1999; 117:670-672.
16. Kyle RA, Shampo MA. Allvar Gullstrand. JAMA 1977; 238:951.
17. Mart'inez MG, Toledo-Pereyra LH. Allvar Gullstrand: Cirujano, Fisico y Premio Nobel. Cir Ciruj 2000; 68:26-31.
18. Carrel A. Man, the Unknown. New York: Harper and Brothers, Pub, 1935.
19. Malinin TT. Surgery and Life: The Extraordinary Career of Alexis Carrel. New York: Harcourt Brace Jovanovich, 1979.
20. Murray J. Human organ transplants: Background and consequences. Surgery 1992; 256:1411-16.
21. Young AB. Scalpel: Men Who Made Surgery. New York: Random House, 1956.
22. Toledo-Pereyra LH. Alexis Carrel: Cientifico, filosofo y cirujano. Cir Gen 1998; 20:246-54.
23. Durkin JT. Hope for Our Time: Alexis Carrel on the Man and Society. New York: Harper and Row, 1965.
24. Jain KM, Swan KG, Casey KG. Nobel prize winners in surgery, Part 2. Am Surg 1982; 48:191-196.
25. Holmgren G. Robert Barany, 1876-1936. Ann Otol 1936; 45:593-595.
26. Shampo MA, Kyle RA. Robert Barany. JAMA 1980; 243:1914.
27. Diamant H. The Nobel prize award to Robert Barany-a controversial decision? Acta Otolaryngol Suppl 1984; 406:1-4.
28. Martínez MG, Toledo-Pereyra LH. Robert Barany: Cirujano, Controversia y Premio Nobel. Cir Ciruj 2000; 68(2):80-85.
29. Stevenson L. Sir Frederick Banting. Toronto: Ryerson Press, 1946.
30. Stevenson LG. Nobel Prize Winners in Medicine or Physiology 1901-1950. New York: Henry Schurman, 1953.
31. Bliss M. The Discovery of Insulin. Chicago: University of Chicago Press, 1982.
32. Jain KM, Swan KG, Casey KG. Nobel prize winners in surgery, Part 3. Am Surg 1982; 48:287-292.

33. Frederick Grant Banting: Codiscoverer of insulin. JAMA 1966; 6:660-661.
34. Morris JB, Schirmer WJ. The "right stuff": Five Nobel prizewinning surgeons. Surgery 1990; 108:71-80.
35. Bliss M. Rewriting medical history: Charles H. Best and the Banting and Best myth. J Hist Med Allied Sci 1993; 48:253-274.
36. Mart'ınez MG, Toledo-Pereyra LH. Frederick Grant Banting: Cirujano, Caballero y Premio Nobel. Cir Ciruj 2000; 68:124.
37. Waser PG. Walter Rudolf Hess: His life and activities at the University of Zurich Medical School Centennial Celebration of his birth: 14 March 1981. Gesnerus 1982; 39:279-286.
38. Akert K. Obituary:Walter Rudolf Hess (1881-1973). Brain Res 1974; 68:V-VIII.
39. Huber A. Walter Rudolf Hess as an ophthalmologist. Gesnerus 1982; 39:287-293.
40. Martínez MG, Toledo-Pereyra LH. Walter Rudolf Hess: Cirujano, Fisiologo ye Premio Nobel. Cir Ciruj 2000; 68:132.
41. Fontenot C, O'Leary JP. Doctor Werner Forssmann's selfexperimentation. Am Surg 1996; 62:514-515.
42. Jain KM, Swan KG, Casey KF. Nobel Prize winners in surgery, Part 4. Am Surg 1982; 48:495-500.
43. Steckelberg JM, Vlietstra RE, Ludwig J et al. Werner Forssmann (1904-1979) and his unusual success story. Mayo Clin Proc 1979; 54:746-748.
44. Martínez MG, Toledo-Pereyra LH. Werner Theodor Otto Forssmann: Cirujano, Cateterista y Premio Nobel. Cir Gen 2000; 22:257.
45. Huggins CB. The business of discovery in the medical sciences. JAMA 1965; 194:1211-1215.
46. Classics in Oncology: Charles Brenton Huggins. CA Cancer J Clin 1972; 22:230-231.
47. Drucker WR. Charles B. Huggins, MD FACS (hon). Surg Forum 1973; 24:V.
48. Huggins CB. On medical investigation. Surg Clin North Am 1969; 49:455-457.
49. Division of Biological Sciences, the University of Chicago. Charles Brenton Huggins, MD: Nobel Laureate Physiology and Medicine 1966. Chicago: University of Chicago Press, 1966.
50. Jurkiewicz MJ. Nobel laureate: Joseph E. Murray, clinical surgeon, scientist, teacher. Arch Surg 1990; 125:1423-1424.
51. Goldwyn RM. Joseph E. Murray, MD, nobelist: Some personal thoughts. Plast Reconst Surg 1991; 87:1110-1112.
52. Merrill JP, Murray JE, Harrison JH et al. Successful homo-transplantations of the human kidney between identical twins. JAMA 1956; 160:277-282.
53. Moore FD. A Nobel Award to Joseph E. Murray, MD: Some historical perspectives. Arch Surg 1992; 127:627-632.

54. Murray JE. Reminiscences for the "50-year retrospective" of transplantation. Transp Proc 1999; 31:34.
55. Schirmer WJ, Morris JB, Dr. Joseph E. Murray: Another hero on the cutting edge. Surgery 1991; 109:806.
56. Tilney NL. The Brigham. The Chimera 1991; 3:16-18.
57. Toledo-Pereyra LH, Palma-Vargas JM. Searching for history in transplantation: Early modern attempts at surgical kidney grafting. Transp Proc 1999; 31:2945-2948.
58. Toledo-Pereyra LH, Toledo AH. 1954. J Invest Surg 2005; 18:285-290.

27

Conclusion

Your interest for surgery, history and humanities is evident, particularly after you finished reading and assimilating all previous works presented in this well-developed and hopefully worthwhile book.

The notes on philosophy, humanities and writing you encountered, the virtues of man you analyzed, the surgeons, pioneers and educators you reviewed, and the date and winners you assessed, trustfully integrated in a comprehensive manner the life and virtues of the practicing surgeon.

In closing, I would like to leave some of the words that were utilized on the *Vignettes on Surgery, History and Humanities*:

"This book presents a glimpse of how surgery evolved through the years, how some of its important representatives responded to the challenges of science and life in general, how their teachings can offer insights and guidance into our way of thinking in our own daily medical practice.

As we present these stories, we would like to leave with the reader the optimism and resourcefulness of these great professionals who represent the best in the surgeon's work and spirit of this medical specialty.

We hope you share with and encourage in future generations of students, residents, and surgical staff an interest and curiosity for understanding those who were at the beginning of this great medical undertaking in the field of surgery and appreciate the meaning of their achievements."

Luis H. Toledo-Pereyra
Editor

Reminiscences on Surgery, History and Humanities,
edited by Luis H. Toledo-Pereyra. ©2007 Landes Bioscience.

Figure 1.   Obtained from Rutkow IM, Zabdiel Boylston and Small pox Inoculation, available at http://archsurg.ama-assn.org/issues/v136n10/ffull/ssh1001-1.html, accessed November 18, 2002.

Figure 2.   Obtained from http://elane.stanford.edu/wilson/Text/4j.html.

Figure 3.   Obtained from Edward Warren, *The Life of John Collins Warren, MD Compiled Chiefly from his Autobiography and Journals*. Boston: Ticknor and Fields, 1860.

Figure 4.   Obtained from http://www.general-anaesthesia.com/people/valentine-mott.html. Accessed December 12, 2005.

Figure 5.   The painting is the property of Jefferson Medical Collage of Thomas Jefferson University in Philadelphia, Pennsylvania (obtained from http://www.jefferson.edu/eakins/images/full_gross.jpg.

Figure 6.   Obtained from the Moorland-Spingarn Research Center, Founders Library, Howard University, Washington, D.C.

Figure 7. . Reproduced with permission from the Chicago Historical Society.

Figure 8.   Obtained from Hugh Young, *A Surgeon's Autobiography*, Harcourt, Brace, and Company, New York, 1940.

Figure 9.   Obtained from Ravitch MM, ed. *The Papers of Alfred Blalock*, Vols. 1 and 2. Baltimore: The Johns Hopkins Press, 1966.

Figure 10.   Obtained from Howard University, Washington, D.C.

Figure 11.   Top: Obtained from Callaghan JJ, Rosenberg AG, Rubash HE, eds. *The Adult Hip*, Vol. 1. Philadelphia: Lippincott-Raven, 1998:23. Bottom: Obtained from Waugh W. *John Charnley: The Man and the Hip*. London: Springer-Verlag, 1990:15,196, 217.

Figure 12.   Obtained from the booklet of the Unveiling of the First Successful Kidney Transplantation at the Countway Library, Boston, MA.

Figure 13.   Obtained from the front cover of the book, *Maestros Nobel de la Cirugia*, published in 2000 by JGH Editores (Mexico).